Your Role in the Green Environment

Trainee Guide
First Edition

Upper Saddle River, New Jersey
Columbus, Ohio

National Center for Construction Education and Research

President: Don Whyte
Director of Product Development: Daniele Stacey
Green Construction Topics Project Manager: Daniele Stacey
Production Manager: Tim Davis
Quality Assurance Coordinator: Debie Ness
Editors: Rob Richardson and Matt Tischler
Desktop Publishing Coordinator: James McKay
Production Assistant: Brittany Ferguson

Development services provided by Topaz Publications, Liverpool, NY
Project Manager: Veronica Westfall
Desktop Publisher: Joanne Hart
Art Director: Megan Paye
Permissions Editors: Andrea LaBarge, Jackie Vidler

Pearson Education, Inc.
Product Manager: Lori Cowen
Product Development Editor: Janet Ryerson
Operations Specialist: Laura Weaver
Cover Design: Kristina D. Holmes and Tim Davis

The cover of this book was printed on Forest Stewardship Council (FSC) approved paper by Phoenix Color Corp. The text pages of this book were printed on FSC Certified Sterling Ultra Digital Stock by Document Technology Resources.

10 9 8 7 6 5 4 3 2 1

ISBN 0-13-602303-7
978-0-13-602303-6

Contren® Learning Series

Preface

TO THE TRAINEE

If you're just getting started on the path towards a career in construction, you'll soon learn about the tools you'll need to carry you forward—from safety awareness, to construction math, to working with hand and power tools. You'll also discover not just how much these skills can help you on the job, but how well they can transfer into your everyday lives. While tackling that next home project, you'll find your new skills will come in handy—enabling you to manage the project more efficiently and safely than you would have before.

The goal of *Your Role in the Green Environment* is to impart the very same work and life-changing skills. You'll not only learn how much of an impact construction has on the environment and ways in which you can lessen the impact, but you'll also learn about the enormous influence you can have on the environment in your day-to-day lives. While these skills may not directly affect how well you accomplish that next home project, you'll likely approach it with a deeper understanding that you didn't have before.

NEW WITH THIS UPDATE TO *YOUR ROLE IN THE GREEN ENVIRONMENT*

This update to Your Role in the Green Environment contains the following enhancements:

- LEED Version 2.2 has been updated to LEED Version 3. The LEED for New Construction Project Checklist is now based on a 100-point model. For more information on the advancements to the new version of LEED, visit the United States Green Building Council at www.usgbc.org.
- Includes a Green Advantage® Study Guide, produced in consultation with Green Advantage® to assist in preparation for the Green Advantage Commercial/Residential Certification Exam. Green Advantage® creates and delivers a nationally recognized, third party green certification for construction practitioners. This certification may earn a LEED Innovation credit for your project. To find out how, contact Green Advantage® at www.greenadvantage.org.
- A list of common acronyms used in the green building environment.
- Updated statistics throughout.
- Revised and updated review and test questions.
- Revised and updated text based on new the LEED rating system.
- "Green Jobs" Initiative Framework chart published by the Employment and Training Administration of the United States Department of Labor. To view any updates to this initiative, visit www.doleta.gov/brg/GreenJobs

SPECIAL FEATURES OF THIS MODULE

In an effort to provide a comprehensive user-friendly training resource, NCCER has incorporated many different features for your use. These features include:

- *Notes, Cautions, and Warnings* – Safety features are set off from the main text in highlighted boxes and are organized into three categories based on the potential danger of the issue being addressed. Notes simply provide additional information on the topic area. Cautions alert you of a danger that does not present potential injury but may cause damage to equipment. Warnings stress a potentially dangerous situation that may cause injury to you or a co-worker.
- *What's Wrong with This Picture?* – What's Wrong with this Picture? features include photos of actual code violations for identification and encourage you to approach each installation with a critical eye.
- *Think About It* – Think About It features use What if? questions to help you apply theory to real-world experiences and put your ideas into action.

- *Going Green* – Going Green looks at ways to preserve the environment, save energy, and make good choices regarding the health of the planet. Through the introduction of new construction practices and products, you will see how the greening of America has already taken root.
- *Trade Terms* – Trade Terms are discussed within the text and defined in the Glossary at the end of the module. These terms are denoted in the text with bold blue type upon their first occurrence.
- *Review Questions* – Review Questions are provided to reinforce the knowledge you have gained. This makes them a useful tool for measuring what you have learned.
- *Your Role in the Green Environment E-Book* – This book is also available in an e-book version through ContrenConnect. Please visit the online catalog at www.nccer.org or www.contrenconnect.com to learn more about acquiring e-access to *Your Role in the Green Environment*.

You are invited to visit the NCCER website at **www.nccer.org** for the latest releases, training information, newsletter, Contren® product catalog, and much more. Your feedback is welcome. You may email your comments to **curriculum@nccer.org** or send general comments and inquiries to **info@nccer.org**.

CONTREN® LEARNING SERIES

The National Center for Construction Education and Research (NCCER) is a not-for-profit 501(c)(3) education foundation established in 1995 by the world's largest and most progressive construction companies and national construction associations. It was founded to address the severe workforce shortage facing the industry and to develop a standardized training process and curricula. Today, NCCER is supported by hundreds of leading construction and maintenance companies, manufacturers, and national associations. The Contren® Learning Series was developed by NCCER in partnership with Pearson Education, Inc., the world's largest educational publisher.

Some features of NCCER's Contren® Learning Series are as follows:

- An industry-proven record of success
- Curricula developed by the industry for the industry
- National standardization providing portability of learned job skills and educational credits
- Compliance with Office of Apprenticeship requirements for related classroom training *(CFR 29:29)*
- Well-illustrated, up-to-date, and practical information

NCCER also maintains a National Registry that provides transcripts, certificates, and wallet cards to individuals who have successfully completed modules of NCCER's Contren® Learning Series. *Training programs must be delivered by an NCCER-Accredited Training Sponsor in order to receive these credentials.*

Foreword

An enormous shift is underway in the U.S. due to two major challenges facing Americans and our economy: rapidly rising energy prices and climate change. The financial pain for most of us is great and is forcing us to rethink how we live, how we get around, where we live, and how we can operate our businesses to cope with this rapidly changing situation. Even more challenging is climate change which threatens higher global temperatures, possibly resulting in crop loss, rising sea levels, more violent storms, and many other negative consequences. Clearly, change is afoot and responding to these twin threats is a national imperative.

Within the construction sector of the U.S. economy there is a significant effort underway to address these challenges. Known simply as *green building*, this movement seeks to design, build, and operate facilities and homes that use far less energy than their conventional counterparts and that reduce the climate change impacts typically associated with buildings. Green buildings also address water consumption and wastewater generation; construction and demolition waste; recycling and reuse of land, buildings, materials, and water; impacts of construction on the building site; and the health of the building.

Systems to support green building, such as the U.S. Green Building Council's LEED® (Leadership in Energy and Environmental Design) green building rating system are emerging and helping accelerate the shift to buildings that minimize energy consumption and that have greatly reduced contributions to climate change. A significant segment of the design community (architects, engineers, interior designers, and landscape architects) has undergone additional training to become LEED Accredited Professionals (LEED-AP) to provide them with a detailed understanding of what constitutes a green building and how to design and deliver them to the marketplace.

The result is that over 20,000 commercial and institutional buildings have been registered with the U.S. Green Building Council for green building certification. A similar phenomenon is happening in residential construction with thousands of homes being certified to local and national green home standards around the country.

Indeed, there is a national program designed specifically for the purpose of educating and certifying contractors, subcontractors and progressive members of the construction community in green building. Known as Green Advantage®, it provides a systematic approach for training about green buildings in the form of a 15-hour course followed by a commercial/residential certification examination. The resulting Green Advantage certification is an indication that the participant has demonstrated his or her knowledge about green building, the vocabulary of green building, and a variety of the unique approaches to construction that are associated with green building.

This volume represents the efforts of the National Center for Construction Education and Research (NCCER) to support Green Advantage training through the NCCER training program sponsor network. This represents a remarkable breakthrough in green building because it means that a potentially enormous audience will now be informed about high-performance building. More importantly it will help embed green building practices in the U.S. construction community thus addressing the issues of climate change and high energy costs.

The efforts of Dr. Annie Pearce at Virginia Tech to bring together a wide and varied array of information into this volume are commendable. This training program marks a major milestone on the road to more responsible construction practices and indicates that the tipping point in green building has probably been reached.

Finally, NCCER's foresight in supporting the development of this volume should be gratefully acknowledged by the larger U.S. green building movement because of the enormous positive impact it will bring in educating contractors, subcontractors, supervisors, and craftspeople alike, in the building of environmentally responsible facilities and homes.

Dr. Charles J. Kibert
Powell Center for Construction and Environment
University of Florida
Gainesville, Florida

Contren® Curricula

NCCER's training programs comprise over 50 construction, maintenance, and pipeline areas and include skills assessments, safety training, and management education.

Boilermaking
Cabinetmaking
Carpentry
Concrete Finishing
Construction Craft Laborer
Construction Technology
Core Curriculum: Introductory Craft Skills
Drywall
Electrical
Electronic Systems Technician
Heating, Ventilating, and Air Conditioning
Heavy Equipment Operations
Highway/Heavy Construction
Hydroblasting
Industrial Coatings
Industrial Maintenance Electrical and Instrumentation Technician
Industrial Maintenance Mechanic
Instrumentation
Insulating
Ironworking
Masonry
Millwright
Mobile Crane Operations
Painting
Painting, Industrial
Pipefitting
Pipelayer
Plumbing
Reinforcing Ironwork
Rigging
Scaffolding
Sheet Metal
Site Layout
Sprinkler Fitting
Welding

Pipeline
Control Center Operations, Liquid
Corrosion Control
Electrical and Instrumentation
Field Operations, Liquid
Field Operations, Gas
Maintenance
Mechanical

Safety
Field Safety
Safety Orientation
Safety Technology

Management
Introductory Skills for the Crew Leader
Project Management
Project Supervision

Spanish Translations
Acabado de Concreto
Aislamiento
Albañilería
Andamios
Carpintería de Formas
Currículo Básico: Habilidades Introductorias del Oficio
Electricidad
Herreria de Refuerzo
Instalación de Rociadores Nivel Uno
Instalación de Tubería Industrial
Orientación de Seguridad
Seguridad de Campo

Supplemental Titles
Applied Construction Math
Careers in Construction
Tools for Success
Your Role in the Green Environment

Acknowledgments

SUSTAINABLE FACILITIES AND INFRASTRUCTURE RESEARCH TEAM

This module was developed by the Sustainable Facilities & Infrastructure Research Team of the Myers-Lawson School of Construction at Virginia Tech. Contributors included Dr. Annie Pearce, Dr. Christine Fiori, and Mr. Sushil Shenoy. To ask questions or learn more about the ideas presented in this module, contact sustainablefacilities@vt.edu.

UNITED STATES GREEN BUILDING COUNCIL®

NCCER is a USGBC Education Provider. NCCER is committed to enhancing the ongoing professional development of building industry professionals and LEED Accredited Professionals (LEED AP) through appropriate Education Courses. As a USGBC Education Provider, NCCER has agreed to abide by USGBC-established operational and educational criteria and is subject to annual reviews and audits for quality assurance purposes. To learn more about the United States Green Building Council, the LEED rating system, or LEED-AP, visit www.usgbc.org.

GREEN ADVANTAGE®

Green Advantage® has approved this module as a good primer for green building concepts and techniques. When taught by qualified instructors, it should assist students and practitioners in their preparations for the Green Advantage® Commercial/Residential certification exam. In addition, a Green Advantage® study guide is included. For more information, contact NCCER customer service; to learn more about Green Advantage® and to better understand the benefits of Green Advantage® certification – visit www.greenadvantage.org.

NCCER PARTNERING ASSOCIATIONS

American Fire Sprinkler Association
Associated Builders and Contractors, Inc.
Associated General Contractors of America
Association for Career and Technical Education
Association for Skilled and Technical Sciences
Carolinas AGC, Inc.
Carolinas Electrical Contractors Association
Center for the Improvement of Construction Management and Processes
Construction Industry Institute
Construction Users Roundtable
Design-Build Institute of America
Green Advantage
Merit Contractors Association of Canada
Merit Shop Training, Inc.
Metal Building Manufacturers Association
NACE International
National Association of Manufacturers
National Association of Minority Contractors
National Association of Women in Construction
National Insulation Association
National Ready Mixed Concrete Association
National Systems Contractors Association
National Technical Honor Society
National Utility Contractors Association
NAWIC Education Foundation
North American Crane Bureau
North American Technician Excellence
Painting and Decorating Contractors of America
Portland Cement Association
SkillsUSA
Steel Erectors Association of America
U.S. Army Corps of Engineers
University of Florida
Women Construction Owners and Executives, USA

Your Role in the Green Environment

Green Goes Platinum

This building, part of the Biodesign Institute at Arizona State University, earned United States Green Building Council (USGBC) Leadership in Energy and Environmental Design (LEED®) Platinum certification for its environmentally friendly design. Design features include a central skylight to provide natural daylight and an automatic louver system that tracks the sun to provide shade from the intense Arizona heat. Other features include a reflective roof and light-colored pavement to minimize heat absorption, the use of native desert landscaping to reduce water use, a rainwater collection system, and a construction waste management plan that reduced landfill waste by over 60 percent.

70101-09

70101-09

Your Role in the Green Environment

Topics to be presented in this module include:

Overview

The construction industry is changing. In this new era, the green environment is an important consideration. As a construction craft worker, you must understand how your daily activities at work and at home affect the green environment. With this knowledge, you can make smart choices to reduce your impact. This module explains how the things you do each day can make a difference. You will learn to measure your carbon footprint and find ways to reduce it. You will also learn how the buildings you construct affect the green environment and how to apply the principles of a green building rating system.

Objectives

When you have completed this module, you will be able to do the following:

1. Describe the major challenges to the green environment that are caused directly or indirectly by the built environment.
2. Identify decisions and actions in your personal and work life that impact the green environment.
3. Prioritize your actions in terms of which ones matter most for the green environment.
4. Describe the life cycle phases of a building and the impacts on the green environment over its life cycle.
5. Identify green alternatives to conventional building practices and describe the pros and cons of those alternatives.
6. Identify specific practices you can implement at work to improve your impacts on the green environment.
7. Describe the Leadership in Energy and Environmental Design (LEED) rating process.
8. Identify construction activities that contribute to a project's LEED rating.
9. Identify common construction pitfalls that may affect a project's LEED rating.

Required Trainee Materials

1. Pencil and paper
2. Calculator

Trade Terms

Trade terms are highlighted in blue. Their definitions can be found in the Glossary at the back of this module.

1.0.0 ◆ INTRODUCTION

The impacts of people on the green environment are considerable. Major concerns include the depletion of resources and **global climate change**. Global climate change, also called global warming, involves changes in weather and temperature worldwide. Your daily activities have an impact on these larger problems. The products you buy, the energy you consume, the waste you throw away, and the travel you undertake all contribute to your overall **carbon footprint**, which is a measure of your contribution to global climate change. Many opportunities exist for you to reduce your carbon footprint, including reducing your overall energy and fuel use; rejecting, reducing, reusing, and **recycling** various products; planting vegetation; and finding better energy sources.

1.1.0 The Nature of Change

Along with growth and change in the green environment, the manmade environment is also changing. Some changes are independent of human actions. The changes in the seasons, for example, are a result of the Earth's angle relative to the sun. These changes have regular and predictable effects on temperature and weather in many parts of the world. *Figure 1* shows the brightly colored leaves that mark the changing of summer to autumn in the northern United States. Seasonal changes are part of what makes the setting for each region of the country unique. Development in each region should be responsive to local climate. It should also reflect the special demands of that climate on the built environment.

Other changes are more severe and less predictable. An example is severe weather patterns that may be associated with the seasons. In the United States, hurricanes are common in summer and fall. Snow and ice storms happen in the winter and tornados are more common in spring and summer. *Figure 2* shows a satellite view of a hurricane, one of the most severe types of weather that affects the United States.

101F01.EPS

Figure 1 ◆ Colored leaves mark the changing of seasons in many regions.

These severe weather events also cause a challenge for the manmade environment. Built facilities must be able to function through severe weather events. The function of these facilities is to preserve the safety and security of their occupants. Designing facilities that can survive severe storms is important.

101F02.EPS

Figure 2 ◆ Satellite view of a hurricane.

1.1.1 Changes in the Manmade Environment

Just as the green environment changes, the manmade environment also changes. The role of the manmade environment is to enable humans to survive and prosper. The built environment also provides shelter from the effects of weather, climate, and other environmental conditions. Over time, the complexity and scale of the built environment has grown. This growth serves to accommodate evolving human needs and desires. Some changes in the manmade environment involve new technologies and practices that make buildings more functional, comfortable, or efficient, while other changes respond to demands for changing styles.

DID YOU KNOW?

Worldwide Power Use is Growing at a Staggering Rate

According to statistics from the U.S. Energy Information Administration, the world's total power usage nearly doubled between 1985 and 2005. In Hong Kong (shown here), power usage during that same period nearly tripled.

101SA01.EPS

Expectations for built facilities have changed as well. The average house size in the United States has grown. According to the National Association of Home Builders, the average house size grew from 1,500 square feet in 1973 to nearly 2,500 square feet in 2007. During the same time, the average family size decreased from 3.14 people per household to 2.57. This means that the amount of house per person has risen nearly 50 percent over the past 25 years.

At the same time, the development of buildings and neighborhoods has had an impact on health and well-being. Studies by the Federal Highway Administration have found that Americans make fewer than 6 percent of daily trips on foot. Adding to this is the fact that much of the development in the United States is designed around the automobile. The absence of sidewalks and bicycle paths in neighborhoods does not encourage people to walk. Large distances between residential areas and services such as stores and schools mean that people are more likely to drive. A 2003 study by researchers at the University of Maryland investigated the health effects of urban **sprawl** (*Figure 3*). They found that people who live in neighborhoods characterized by urban sprawl weigh about six pounds more than people living in compact neighborhoods. Compact neighborhoods have sidewalks and shops close to residential areas. This evolution toward less compact neighborhoods has implications on the green environment as well.

1.1.2 Relationships between Human Activities and the Green Environment

As human populations grow and expectations for standards of living increase, there is a greater demand for resources from the green environment. The **raw materials** used to build, furnish, and operate buildings are mined, harvested, or extracted from the environment. The fuel required to power cars, factories, and buildings is also mined or extracted from the environment. The burning of these fuels has consequences for the green environment as well.

The impacts of human activities on the green environment are seen at many levels. You can see local impacts in your own backyard. There are also global impacts whose effects may not be seen for many years (*Figure 4*). Either way, the impacts of human beings on the green environment are undeniable. Major environmental challenges to which human activities contribute include:

101F03.EPS

Figure 3 ◆ Urban sprawl leads to increased transportation requirements.

Figure 4 ◆ The impacts of human activities on the green environment.

- *Global climate change* – The increase in greenhouse gases is producing an overall rise in global temperatures. These gases are due in part to the burning of fossil fuels as a result of **urbanization**. Urbanization involves people moving to the suburbs, which results in longer commutes. Temperature increases affect sea levels as reserves of ice at the Earth's poles melt. They also increase the number of severe weather events.
- *Excess wood harvesting* – The harvesting of wood resources at an unsustainable rate is known as **deforestation**. It leads to soil depletion, pollution of streams, and habitat loss. Deforestation also contributes to global warming because trees reduce greenhouse gases. **Preservation** of forest areas can help to slow global warming.
- *Species extinction* – Habitat loss due to human development has caused thousands of species to become extinct. This is known as a loss of **biodiversity**. The remaining habitats are fragmented and degraded in quality. Human beings depend on diverse **ecosystems** to purify the air and water. These ecosystems help to stabilize climate change. They also provide a variety of resources from lumber to medicine. Ecosystem health is essential for human survival.
- *Decreasing water supplies* – The rise in global temperatures has caused **desertification** (the spread of deserts) in dryer parts of the world. In other areas, overgrazing and the overuse of groundwater has led to **aquifer depletion**. This is particularly true in agricultural and urban areas. **Aquifers** are the only reliable source of water in many parts of the country.
- *Air and water pollution* – Industrial activities and the daily activities of people contribute to the pollution of air and water. The burning of fossil fuels for transportation and energy produces smog, which leads to **acidification** (acid rain) and plant decline. The use of these fuels also contributes to ground level ozone and other forms of air pollution. Runoff from paved and deforested areas contributes to water pollution. Industrial deposits into waterways increase water temperature and contaminant levels. Many modern components of wastewater cannot be removed using normal treatment methods. These include such things as prescription drugs and plastic residuals, which accumulate in watersheds with unknown long-term consequences for wildlife.

- *Soil contamination and depletion* – Contaminants from human activities often remain in the soil and eventually migrate to water sources. Sources of contamination range from leaking underground storage tanks to **stormwater runoff** from paved areas. Removing vegetation exposes topsoil and causes **erosion** losses.
- *Loss of ozone* – Release of chlorine-based contaminants such as refrigerants has led to **ozone depletion** in the upper atmosphere. Ozone molecules are necessary to shield living organisms from **solar** radiation. Reduced levels of ozone are thought to be responsible for the increased rate in skin cancer and cataracts as well as damage to marine and terrestrial ecosystems. These chemicals also contribute to a greater-than-usual growth in the seasonal **ozone hole** that appears naturally in the polar regions.

None of these occur in isolation. In many cases, human activities contribute to more than one problem at a time. All of them require attention to ensure the health of the planet for future generations.

1.2.0 Impact of Individual Human Activities

Given the growing awareness of human impacts on the green environment, how do your daily activities contribute to environmental problems and their solutions? The manufacturing, transportation, use, and disposal of the products used in daily living contribute to each person's individual impacts. Your individual carbon footprint can serve as a measure of how your lifestyle contributes to global climate change. Your work on the job site and the buildings you help to create also contribute to the challenges being faced by the green environment.

1.2.1 The Average American Household

The average U.S. household is annually responsible for the production of 3,500 pounds of garbage, 450,000 gallons of wastewater, and 54,600 pounds of carbon dioxide (CO_2), along with smaller amounts of sulfur dioxide (SO_2), nitrogen oxides (NO_X), and heavy metals. These impacts come from the use of products, the use of resources such as electricity, the types of waste generated, and travel (*Figure 5*).

The Consumer Electronics Association estimates that the average household spends nearly $1,200 per year on electronics, including televisions, digital cameras, and other devices. According to the Environmental Protection Agency (EPA), the average household also spends about $1,900 per year to run them, along with all the other uses of energy in the home such as heating and lighting. In fact, according to a 2006 study conducted by Nielson Media Research, the average American household has more televisions than it does people. In addition to durable products, the U.S. Department of Agriculture reports that Americans also spend an average of $40 per person per week on food, much of which requires extensive use of water and fossil fuel energy to produce, process, and transport to the table.

101F05.EPS

Figure 5 ◆ Your individual choices make a difference.

According to the EPA, the average household in the United States spends nearly $500 per year on its water and sewer bill. Each person in the U.S. consumes 100 gallons per day on average and generates over 50 gallons of wastewater. (About half of the water consumed is for irrigation and goes back into the ground rather than the wastewater stream.) Of this wastewater, 9,000 gallons per year is used to flush away only 230 gallons of waste from toilets, a surprisingly wasteful use of water that has been treated to drinking water standards. Water use also requires energy for pumping, collection, treatment, and distribution. For example, letting a faucet run for five minutes uses as much energy as leaving a 60-watt light bulb on for 14 hours.

People also produce tremendous amounts of solid waste, both directly in homes and indirectly through the production of consumer goods. The average American generates 4.5 pounds of waste per day, of which 75 percent could be recycled, but only 20 percent typically is. That adds up to just over 1,300 pounds per person per year. On average, each American consumes 3 gallons of gas per day—enough to fill up 21 bathtubs per year. This doesn't count all the trips on public transit, trains, buses, or airplanes.

All this adds up to a considerable impact on the green environment. To determine exactly how much impact you and your family have, complete the inventory of household impacts found in *Worksheet 1*.

WORKSHEET 1: INVENTORY YOUR HOUSEHOLD IMPACTS

Conducting an inventory of your household consumption, waste generation, and activities is the first step in understanding how you can reduce your impact. Answer the following questions based on your best guess. If you share a household with other people, divide the total answer for your household by the number of people who live in your home.

How many gallons of garbage do you put out each week? An average garbage can holds 32 gallons. Multiply by 52 to calculate the gallons of garbage you throw away per year.

Gallons of garbage per year: ______________________

How much electricity do you use per year? You'll find this information on your electricity bill. You can add up the total for 12 months worth of bills, or multiply a monthly average by 12. If you know how much your electricity costs per month, you can estimate the amount of energy used in kilowatt-hours by dividing your total bill by 10.

Total electricity per year in kilowatt-hours: ______________________

How many therms of natural gas do you use per year? Check your natural gas bill if you get one. It will tell you how many therms of gas you use per month. Your highest values will probably be during the winter heating season. If you know your monthly average, multiply by 12 to calculate your annual use.

Total therms of natural gas per year: ______________________

How many gallons of propane do you use each year? You can check this on your propane bill if you get one. Add up the total number of gallons per year.

Total gallons of propane per year: ______________________

How many gallons of fuel oil do you use per year? You can check this on your fuel oil bill if you get one. Add up the total number of gallons per year.

Total gallons of fuel oil per year: ______________________

On average, what is your monthly combined water and sewage bill? Check your monthly bill if you get one and add up all the amounts for a one-year period. If you don't get a water and sewage bill, you can estimate this amount as approximately $75 per person per year.

Annual cost for water and sewage: ______________________

How many square feet is your house or dwelling? If you don't know, draw a floor plan of your house or apartment and estimate the floor area in square feet.

Size of house (floor area) in square feet: ______________________

101WS01A.EPS

WORKSHEET 1 (Continued)

On average, how many miles do you drive your household vehicles per week, and what are their average fuel efficiencies? 300 miles per week per vehicle or 15,000 miles per year is about average in the United States. If you're not sure about fuel efficiency, assume your car gets 22 miles to the gallon, which is about average. If you know how many gallons of fuel you use per week, skip directly to the end of the line.

Car 1 miles per week ___________ / *Car 1 miles per gallon* ___________ = *Car 1 gallons per week* ___________

Car 2 miles per week ___________ / *Car 2 miles per gallon* ___________ = *Car 2 gallons per week* ___________

Car 3 miles per week ___________ / *Car 3 miles per gallon* ___________ = *Car 3 gallons per week* ___________

Add up the gallons of gas per week for all your vehicles, and then multiply by 52 to estimate the gallons of gasoline you use per year.

Total gallons of gasoline you use per year: ___________________________

On average, how much do you travel each year on airplanes? Estimate the number of flight segments below for each of the three distances. A round trip counts as two segments, and each segment of a multi-segment flight counts as its own flight. For example, if you fly from Baltimore to Cincinnati to Atlanta, that counts as two flight segments.

Number of short-haul flight segments (less than 700 miles or 2 hours): ______

Number of medium-haul flight segments (700 – 2,500 miles or 2 – 4 hours): ______

Number of long-haul flight segments (more than 2,500 miles or longer than 4 hours): ______

On average, how many miles do you travel on public transportation per year?

Number of miles per year on transit bus/subway: ______

Number of miles per year on intercity bus: ______

Number of miles per year on intercity train: ______

Add up the total number of miles per year on public transportation: ______________________

Your answers to these questions will help you calculate your own carbon footprint later in the module. Keep track of your answers in the workbook or on a separate worksheet.

101WS01B.EPS

1.2.2 The Impacts of the Products You Use

Every product you use has a hidden history of harvesting, extraction, manufacturing, and transportation that cause impacts beyond what you see in the product itself. The total energy required to bring a product to market is known as **embodied energy**. It includes everything from raw material extraction to manufacturing to final transport and installation. For example, aluminum beverage cans may be manufactured with metals mined in several different countries and then shipped on pallets made from wood harvested in another country, using fuel from yet another country. The can itself is more costly and complicated to manufacture than the beverage. Drinking the beverage takes a few minutes, throwing the can away takes a second, and yet the true impact is immense. Think of all the products you consume or use over the course of a given year. How much of an impact do you think those products have? Answer the questions in *Worksheet 2* to develop an inventory of the products you consume and begin to estimate the impacts those products have. Your answers to these questions will help you calculate your carbon footprint later in this module.

WORKSHEET 2: INVENTORY YOUR PRODUCT IMPACTS

Consider the products you buy or that are bought for you over the course of a year. Answer the following questions based on your best guess. If you share a household with other people, divide the total answer for your household by the number of people who live in your home.

Eating out: $ ________ *per month* × 12 = $ ________ *per year*
Meat, fish, & protein: $ ________ *per month* × 12 = $ ________ *per year*
Cereals & baked goods: $ ________ *per month* × 12 = $ ________ *per year*
Dairy: $ ________ *per month* × 12 = $ ________ *per year*
Fruits & vegetables: $ ________ *per month* × 12 = $ ________ *per year*
Other: $ ________ *per month* × 12 = $ ________ *per year*

On average each month, how much do you spend on the following goods and services? Multiply each amount by 12 to estimate your average annual spending in each category. If you already know how much you spend per year, you can skip the monthly amount and write the average amount per year at the end of each line.

Clothing: $ ________ *per month* × 12 = $ ________ *per year*
Furnishings & household items: $ ________ *per month* × 12 = $ ________ *per year*
Other goods: $ ________ *per month* × 12 = $ ________ *per year*
Services: $ ________ *per month* × 12 = $ ________ *per year*

101WS02.EPS

DID YOU KNOW?

Aluminum Cans

The United States still gets three-fifths of its aluminum from virgin ore, at twenty times the energy intensity of recycled aluminum, and throws away enough aluminum to replace its entire commercial aircraft fleet every three months.

101SA02.EPS

1.2.3 Your Carbon Footprint and Global Climate Change

Greenhouse gases, such as CO_2, create what is known as the Greenhouse Effect (*Figure 6*). A certain level of greenhouse gas is essential because it prevents the loss of heat into outer space. However, if the level increases too much, it can result in global warming. The **carbon cycle** is the transfer of carbon (mostly in the form of CO_2) between Earth and the atmosphere. Plants absorb CO_2 and release oxygen. People (and many of their energy-using activities, such as combustion) absorb oxygen and release CO_2. This cycle has become out of balance as people generate more and more CO_2 while stripping the planet of the trees that absorb CO_2 and generate oxygen.

The inventories you created in the previous section are useful for estimating your own contribution to the Greenhouse Effect, which ultimately leads to global climate change. Any changes made now can help to lessen the severity of these impacts.

For example, the rise in global temperatures is expected to cause a certain portion of polar ice caps to melt, which will cause a rise in sea levels. If you live in a coastal area at a low elevation, you might find that the land around you is disappearing as the ocean level rises (*Figure 7*). At a minimum, you can expect severe weather to occur more frequently—watch for stronger, more frequent hurricanes, tornados, and storms interspersed with

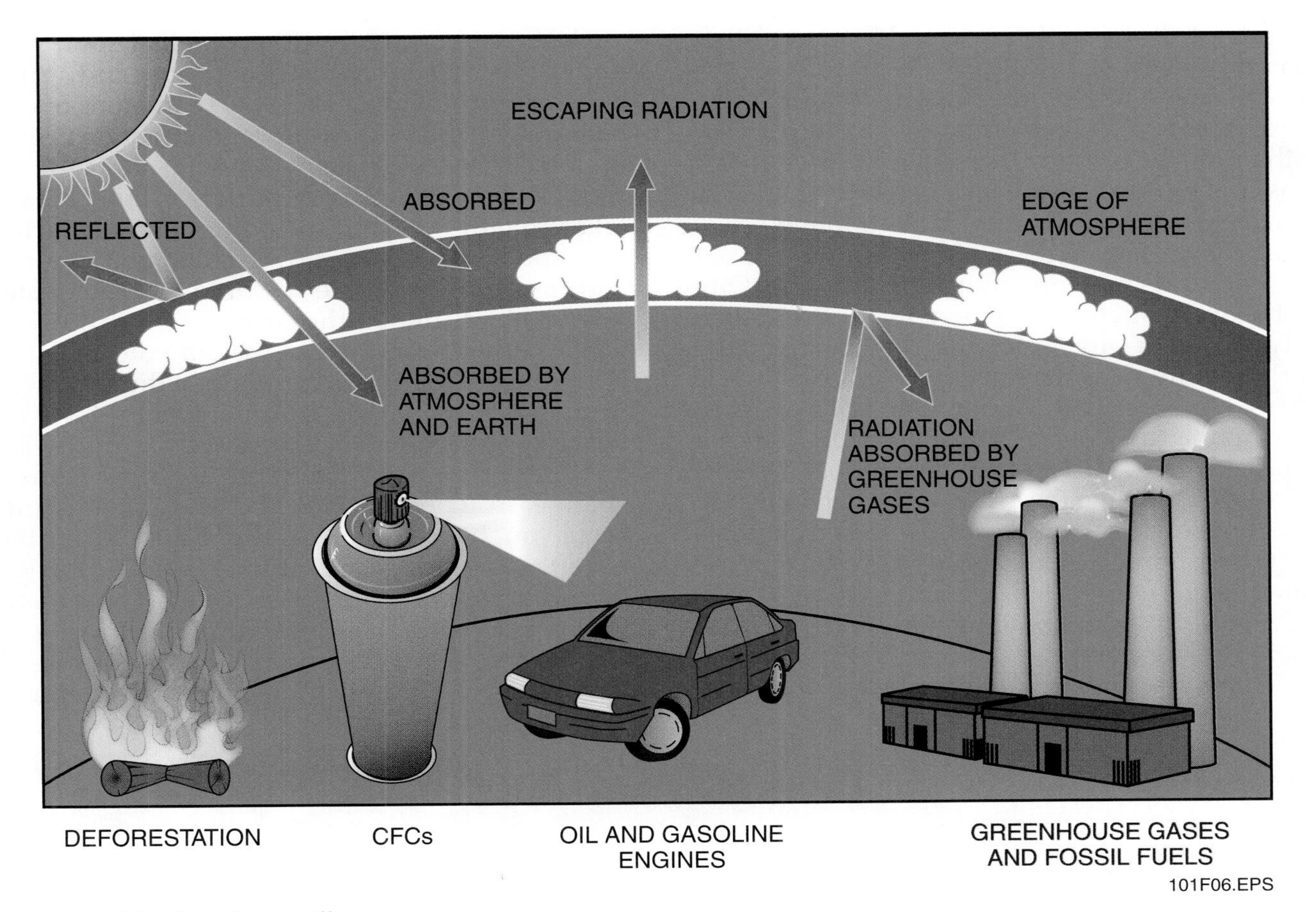

Figure 6 ◆ The Greenhouse Effect.

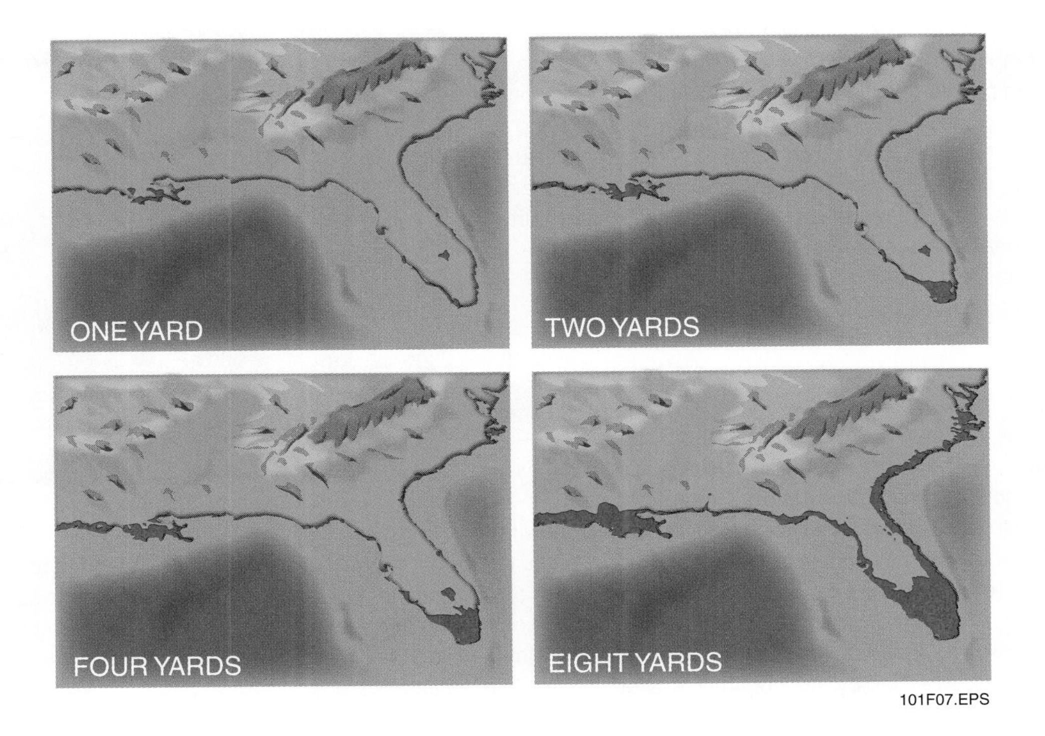

Figure 7 ◆ The effects of rising sea levels.

periods of drought. There are also many other potential effects of global climate change that scientists believe might occur as temperatures rise. These include many possibilities, from an increase in the number of forest fires due to drier conditions to the possible extinction of species such as polar bears due to a loss in habitat. Many scientists believe that increases in greenhouse gases will lead to rising temperatures worldwide. Others believe that climate change may have drastically different effects in different parts of the world—some areas may get much warmer, while others may get much colder. One scientist even refers to the concept as "global weirding" since it is difficult to predict exactly how global systems will respond. While the actual effects of increased greenhouse gases remain to be seen, scientists agree that human activities are resulting in climate change.

One way to evaluate your personal impact on global climate change is through footprint analysis. There are three different types of footprint analysis: carbon footprint, **ecological footprint**, and **water efficiency**/water footprint. Your carbon footprint evaluates how many tons of carbon are emitted as a result of your actions. Your ecological footprint is measured in acres and represents the amount of productive land area it takes to support you, including manufacturing the products you buy, growing the food you eat, and absorbing the waste you produce. Your water footprint estimates the number of gallons of water per year required to sustain your lifestyle. Each of these methods focuses on a different way to calculate the effects you personally have on the green environment. Your carbon footprint is a measure of your contribution to global climate change. Your ecological and water footprints focus on how much of the Earth's limited resources you consume. In fact, your ecological footprint is often expressed in terms of how many Earths it would take if everyone lived like you. The average American's ecological footprint is 30 acres, compared to a world average of 6 acres.

DID YOU KNOW?

Surprising Impacts of Global Warming

In addition to increasing temperatures, severe weather, and rising sea levels, other surprising things that may happen because of global climate change include the following:

POISON IVY

101SA03.EPS

- Rising levels of CO_2 and warmer temperatures are prompting many plants to bloom earlier and produce more pollen, resulting in higher incidence of allergies in many parts of the U.S. Poison ivy, an allergy-causing plant, has been measured to have more potent itch-causing oils when grown in environments with higher CO_2 levels, and tends to thrive in warmer temperatures.
- Increased temperatures are causing changes in habitats that may be prompting animals such as squirrels, mice, and chipmunks to move to higher elevations. These changes also affect other species such as polar bears, whose habitat is threatened as polar ice caps melt away.
- Thawing of permafrost can affect the stability of surfaces and structures on top of them, causing sinkholes, damage to structures, rockslides, and mudslides. Some scientists are concerned that thawing of permafrost may also reveal bodies of animals and people formerly frozen in the ice. As these corpses are discovered, they may be a source of diseases such as smallpox to which humanity is no longer resistant.
- Due to changes in blooming schedules for many plant species and earlier arrival of spring temperatures, animals that migrate based on the number of hours of daylight may find that the food sources on which they depend are no longer available by the time they arrive in their normal migration areas.

(Source: www.livescience.com)

Footprint Calculators

To calculate your ecological footprint, go to www.myfootprint.org. An online calculator for carbon footprint is available at www.carbonfootprint.com. You can calculate your water footprint at www.waterfootprint.org. To see how much your own footprint equals in terms of impacts, go to www.epa.gov/cleanenergy to convert your footprint to equivalents such as tanker trucks' worth of gasoline or railcars full of coal.

Unfortunately, if only 12 percent of the Earth's biosphere or viable surface area is reserved for other species, there are less than 5 acres available per person. With continued population growth and increase in standards of living, the amount of overshoot stands to increase even further. Right now, 20 percent of the Earth's viable ecological footprint is consumed by 2.5 percent of its population—the richest 2.5 percent. Many of the remaining population do not have enough for basic survival. Some cities are making concerted efforts to reduce their ecological footprints (*Figure 8*).

Using the information collected earlier in this module about household activities and consumption, you can estimate how many pounds of carbon per year are generated by each of these activities. For each of the sections in *Worksheet 3*, fill in the amounts from your earlier inventories and multiply by the factors shown to estimate the total pounds of carbon for each activity. Then add up the totals for each category to estimate your total carbon footprint in pounds.

The average American household generates about 54,600 pounds of carbon per year or about 20,000 pounds per person, compared to 27,700 pounds by the average German household, 10,600 pounds by the average Mexican household, and 400 pounds by the average Kenyan household.

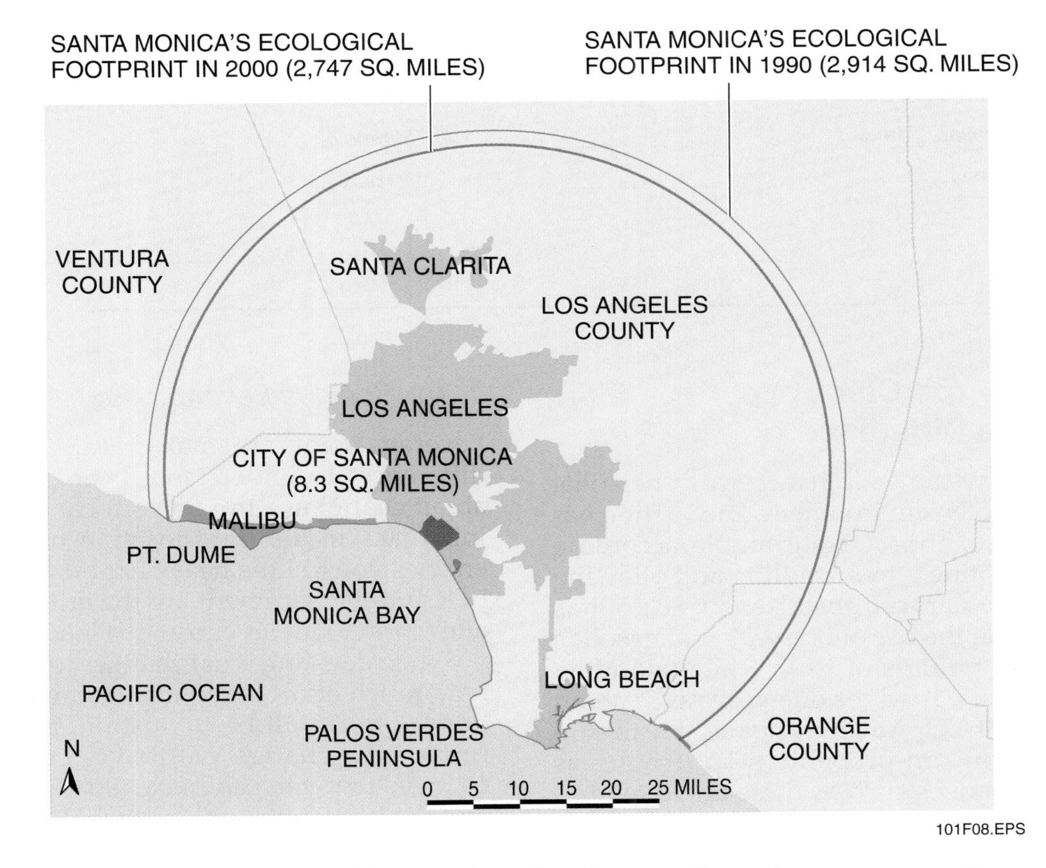

Figure 8 ◆ Santa Monica reduced its ecological footprint through various recycling and energy conservation programs.

WORKSHEET 3: DETERMINE YOUR CARBON FOOTPRINT

Category	Amount	Carbon Factor	Total Pounds of CO_2/Year
Gallons of garbage/year:	________	*× 2 lbs/gallon =*	________*lbs/year*
Total electricity/year in kilowatt-hours:	________	*× 1.4 lbs/kWh =*	________*lbs/year*
Total therms of natural gas/year:	________	*× 11.7 lbs/gallon =*	________*lbs/year*
Total gallons of propane/year:	________	*× 12.7 lbs/gallon =*	________*lbs/year*
Total gallons of fuel oil/year:	________	*× 22.4 lbs/gallon =*	________*lbs/year*
Annual cost for water and sewage:	________	*× 8.9 lbs/$ =*	________*lbs/year*
House size (floor area) in square feet:	________	*× 2.1 lbs/sq ft =*	________*lbs/year*
Gallons of gasoline/year:	________	*× 20 lbs/gallon =*	________*lbs/year*
No. of short-haul flight segments/year:	________	*× 304 lbs/segment (avg) =*	________*lbs/year*
No. of medium-haul flight segments/year:	________	*× 726 lbs/segment (avg) =*	________*lbs/year*
No. of long-haul flight segments/year:	________	*× 2,217 lbs/segment (avg) =*	________*lbs/year*
Miles per year on public transportation:	________	*× 0.5 lb/mile =*	________*lbs/year*
Eating out (dollars/year):	________	*× 0.8 lb/dollar =*	________*lbs/year*
Meat, fish, & protein (dollars/year):	________	*× 3.2 lbs/dollar =*	________*lbs/year*
Cereals & baked goods (dollars/year):	________	*× 1.6 lbs/dollar =*	________*lbs/year*
Dairy (dollars/year):	________	*× 4.2 lbs/dollar =*	________*lbs/year*
Fruits & vegetables (dollars/year):	________	*× 2.6 lbs/dollar =*	________*lbs/year*
Other (dollars/year):	________	*× 1.0 lb/dollar =*	________*lbs/year*
Clothing (dollars/year):	________	*× 1.0 lb/dollar =*	________*lbs/year*
Household items (dollars/year):	________	*× 1.0 lb/dollar =*	________*lbs/year*
Other goods (dollars/year):	________	*× 0.75 lb/dollar =*	________*lbs/year*
Services (dollars/year):	________	*× 0.4 lb/dollar =*	________*lbs/year*
Total =			________________**lbs/year**

101WS03.EPS

1.3.0 Things You Can Do to Make a Difference

You can do many things to reduce your personal impact on the green environment. Each action has a different level of impact. For example, according to the EPA, planting a tree seedling and allowing it to grow for 10 years can absorb about 86 pounds of CO_2. Avoiding the use of one gallon of gasoline saves about 19 pounds of CO_2, while taking an average car off the road for one year saves about 11,765 pounds of CO_2. You may not be able to stop driving, but you can probably stop driving as much. Small changes such as these are known as leverage points.

1.3.1 Seeking Leverage Points

Leverage points are small changes that make a big difference. To understand leverage points, think about changing a tire on your car (*Figure 9*). Placing a jack in the right location under the frame enables you to lift a heavy car off the ground. The jack and its placement are the critical factors that allow you to lift an enormous load.

When deciding what changes to make, consider how much effort is required to make the change, how easy it will be to sustain, and how much impact the change will have. Look for simple changes that you can make (such as using a jack) and the best opportunities to make them that will have the most impact (such as selecting the right place on the car to position the jack). Changes that require you to adjust how you behave on an ongoing basis are often more difficult than changes requiring a single effort.

101F09.EPS

Figure 9 ◆ Leverage points are like a carefully placed car jack.

Finally, be sure to consider all the associated costs and benefits of potential actions, not just the initial costs of purchase. Sometimes the **life cycle** costs of an item make it worthwhile to spend more money up front to achieve greater savings over time. **Life cycle analysis** involves looking at the environmental impact throughout a product's entire life cycle. For example, an energy-efficient light bulb may cost more initially but will pay for itself through reduced energy use. It may also save money through reduced maintenance costs because it lasts longer and therefore won't have to be replaced as often, and it may reduce air conditioning costs because it runs cooler. The amount of time required to save the additional money invested up front is known as the **payback period**.

1.3.2 Reducing Energy Use

One important way to reduce your carbon footprint is by reducing your overall energy use. About 87 percent of all energy used in the United States comes from burning fossil fuels, including oil, coal, natural gas, and propane (*Figure 10*). In fact, the burning of fossil fuels is the largest single

Living Off the Grid

Suppose you want to build a small house in the country. The house will provide a retreat and will have all the comforts of home, but with its rural location, it is too far from existing power supplies to be connected to the grid. How will you provide power for your house and its amenities? Which electrical devices are you willing to live without in order to save energy?

101SA04.EPS

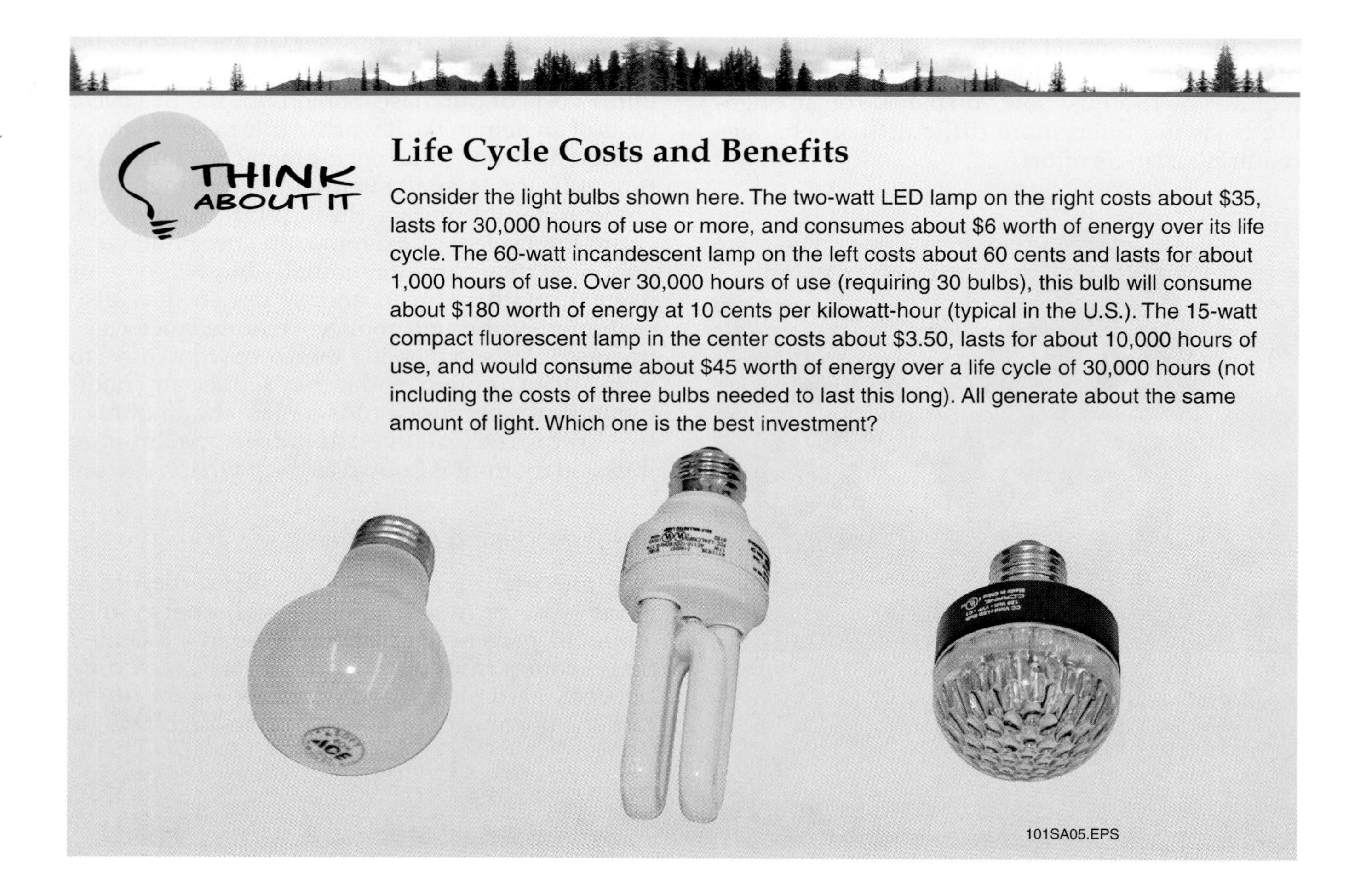

Life Cycle Costs and Benefits

Consider the light bulbs shown here. The two-watt LED lamp on the right costs about $35, lasts for 30,000 hours of use or more, and consumes about $6 worth of energy over its life cycle. The 60-watt incandescent lamp on the left costs about 60 cents and lasts for about 1,000 hours of use. Over 30,000 hours of use (requiring 30 bulbs), this bulb will consume about $180 worth of energy at 10 cents per kilowatt-hour (typical in the U.S.). The 15-watt compact fluorescent lamp in the center costs about $3.50, lasts for about 10,000 hours of use, and would consume about $45 worth of energy over a life cycle of 30,000 hours (not including the costs of three bulbs needed to last this long). All generate about the same amount of light. Which one is the best investment?

101SA05.EPS

U.S. ENERGY CONSUMPTION BY SOURCE

BIOMASS 2.9%
RENEWABLE
HEATING, ELECTRICITY, TRANSPORTATION

PETROLEUM 38.1%
NONRENEWABLE
TRANSPORTATION, MANUFACTURING

HYDROPOWER 2.7%
RENEWABLE
ELECTRICITY

NATURAL GAS 22.9%
NONRENEWABLE
HEATING, MANUFACTURING, ELECTRICITY

GEOTHERMAL 0.3%
RENEWABLE
HEATING, ELECTRICITY

COAL 23.2%
NONRENEWABLE
ELECTRICITY, MANUFACTURING

WIND 0.1%
RENEWABLE
ELECTRICITY

URANIUM 8.1%
NONRENEWABLE
ELECTRICITY

SOLAR & OTHER 0.1%
RENEWABLE
LIGHT, HEATING, ELECTRICITY

PROPANE 1.7%
NONRENEWABLE
MANUFACTURING, HEATING

101F10.EPS

Figure 10 ◆ U.S. energy consumption by source (U.S. Department of Energy).

contributor to carbon footprint in the United States, and it is at the root of many of the challenges faced by the green environment.

The average U.S. household spends its energy dollar as shown in *Figure 11.* It is not surprising that most of the money goes toward heating, cooling, hot water, appliances, and lighting. What may surprise you is the large miscellaneous category. This category includes the electrical devices not specifically listed elsewhere, such as the growing number of battery chargers used to power cellular phones, portable tools, laptop computers, and so on. These chargers draw a small amount of power even when the battery is fully charged. The warmth you feel when you touch a battery charger is some of the wasted power being dissipated as heat.

Look around—how many items plugged into your walls have a clock, timer, or charger built in? All of these items draw what are called phantom loads. In other words, they are consuming energy even when it appears that the device itself is turned off. The only way to stop the power consumption by these devices is to unplug them completely. There are many ways to reduce your use of energy. Some of the methods recommended by GreenerChoices.org are listed below.

To reduce the amount of energy used to heat and cool your home:

- Choose energy-efficient furnaces or air conditioners that are the right size for your home. Check for rebates from your local utility that can help pay to replace your current systems with more efficient ones.
- Properly insulate your home's attic, walls, and slab or crawl space, including all ducts outside the conditioned space of your home.
- Contact your utility company for a free energy audit, or go online to http://hes.lbl.gov for a web-based energy audit tool that can help identify areas where you can improve household energy use.
- Install programmable thermostats, insulated windows, and ceiling fans to increase comfort in your home. Programmable thermostats can save up to 20 percent of your heating and cooling costs by cutting back your systems at night and during the day when you are not at home.

To save energy used for hot water heating:

- Consider replacing your tank hot water heater with a tankless or instantaneous hot water heater that only heats water as needed, or a solar hot water heater if your home design can accommodate one.
- Set your water heater to 120°F instead of 140°F. Most appliances that require hotter water, such as dishwashers, have built-in heating coils.
- Insulate your hot water pipes and hot water tank to reduce heat loss.
- Replace faucets and showerheads with low-flow models if they are more than 10 years old. Newer showerheads can deliver a pleasant shower at 1.5 gallons per minute or less due to the introduction of air into the water stream.

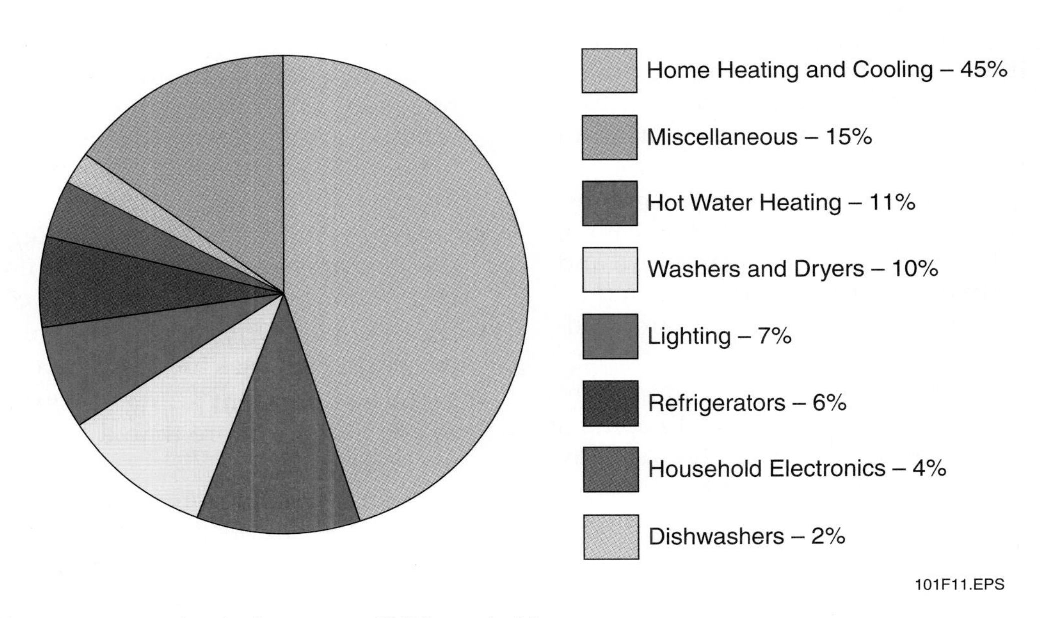

Figure 11 ◆ Energy consumption in the average U.S. household.

To save energy used by appliances such as washers, dryers, refrigerators, and dishwashers:

- Wash your laundry in cold water. It helps your clothes last longer and fade less, and can save over $60 per year in hot water heating costs.
- Consider a front-loading washing machine. These machines save up to 25 percent of the water used for washing, along with the energy required to heat that water. They also use less detergent and remove more water from clothes during the spin cycle, reducing drying time and energy use.
- Use a clothesline instead of a dryer.
- Choose a top-freezer model of refrigerator instead of a side-by-side model. Top-freezer models are more efficient due to the placement of the compressor. Beware of built-in icemakers—they add to the energy cost and are the culprit for many repair calls.
- Keep your appliances full. Full refrigerators maintain more even temperatures and cycle on less frequently. Full washing machines and dishwashers get the maximum amount of cleaning possible for the least use of resources. Load washing machines, dryers, and dishwashers according to the manufacturer's instructions to increase effectiveness.

To save energy in lighting, household electronics, and miscellaneous areas:

- Swap your incandescent bulbs for **compact fluorescent lamps (CFLs)** or LEDs. They use less energy and last longer than a standard incandescent bulb.
- Use power strips and sensors. Power strips can be used to effectively unplug your electronic devices without disrupting cords and cables. Electric photocells and occupancy sensors turn lights off automatically when they are not needed.
- Buy **Energy Star®** appliances and devices. These products perform in the top 25 percent of their product class in terms of energy use and contribute to overall energy **conservation** (*Figure 12*). Listings of approved products are available online at www.energystar.gov.
- Use the sleep mode or shut down equipment such as computers when not in use. Keeping a computer and monitor running 24 hours a day consumes up to 1,100 kilowatt-hours of energy, which could cost over $100 per year depending on electricity rates.

101F12.EPS

Figure 12 ◆ Energy Star® products meet strict requirements for energy conservation and efficiency.

1.3.3 Reducing Fuel Use and Increasing Fuel Efficiency

Over half of the carbon footprint of most Americans is due to transportation choices. Some ways to reduce the overall use of fuel and increase fuel efficiency include the following:

- Drive a fuel-efficient, **low-emission vehicle**. Many options are available, including gasoline-electric **hybrid vehicles** that get over 50 miles per gallon. You may also consider driving a scooter or motorcycle. While most people can't get by with this type of vehicle alone, the fuel efficiency (up to 100 miles per gallon or more) can make scooter commuting a worthwhile investment, particularly if you live in a warmer climate.
- **Carpool** and/or combine trips whenever possible. The greatest amount of air pollution is released when an engine is cold. Combine errands on a single trip whenever possible. This helps vehicles run more efficiently and reduces the overall mileage.
- Maintain your vehicle properly. Regularly check the tire pressure and keep the engine tuned for optimal performance and fuel economy.
- Drive slower. Driving at speeds above 60 miles per hour decreases fuel efficiency.
- Reduce **equipment idling**. If you will be sitting in your car for more than a few minutes, turn the engine off.
- Use mass transportation instead of individual vehicles whenever possible. The fuel used per person is much lower for mass transportation than for travel in a single occupancy vehicle.

- Avoid unnecessary air travel. Airplanes are among the largest contributors to carbon emissions, and their emissions are made worse by the fact that they occur high in the atmosphere. If possible, replace short-haul flights with travel by train or bus to reduce overall impacts.
- Purchase locally produced goods and products wherever possible. One particular area of focus is food. Many foods, particularly fruits and vegetables, are transported thousands of miles before they reach local stores. Shop at local farmers' markets or ask your grocer to stock locally produced foods to minimize the fuel used to transport goods.
- Move information, not people or things. Take advantage of the telephone and internet as much as possible to avoid unnecessary travel. For example, rather than driving to several stores to find out which one carries a product or to compare prices, call the stores instead or research it on the internet.

1.3.4 Rejecting, Reducing, Reusing, and Recycling Materials

The three Rs (reduce, reuse, and recycle) is a common phrase used to remember how to make greener choices. Reducing the materials used has a greater impact than reusing products or recycling them. This is true because using less of a material means less has to be produced, causing less impact from harvesting, manufacturing, and transportation. Likewise, reusing materials is better than using disposable materials that you recycle. Each reuse is one less product that has to be made. When you do use disposable goods, always try to recycle them instead of just throwing them out. Recycling is a good source of raw materials for new products. It is also an important way to save resources for the future.

The best choice of all, however, is a fourth R—reject. Be on the lookout for ways to reject the need to use raw materials in the first place. Rejecting the offer of a bag when you've only purchased one item is a good example. It's one less bag consumed and one less thing to throw away. Most receipts can also be rejected—think twice before you say yes at the gas pump or ATM. Are you really going to use that receipt for anything, or just crumple it up and throw it away? From online statements for your bills to email instead of letters, there are many opportunities to reject the use of raw materials. Together, the four Rs—reject, reduce, reuse, and recycle—can help you remember the best way to be green in the choices you make every day.

1.3.5 Planting Vegetation

One important way to offset carbon emissions is by planting vegetation, such as trees or a garden. As mentioned earlier, plants absorb CO_2 and produce oxygen. They also provide shade, reduce stormwater runoff, promote groundwater recharge, and prevent soil erosion. There are many options for creating a beautiful and functional landscape while minimizing the impact of fertilizers, pesticides, herbicides, and emissions from lawn equipment. These include the following:

- Plant native plants in your landscape instead of alien or invasive species. Native plants are well-adapted to local climate conditions and pests, and can often be grown without irrigation, pesticides, or fertilizer.
- Group plants with similar needs together in the landscape. This allows you to focus the application of irrigation, pesticides, and fertilizer only on the areas that really need it, while avoiding over-irrigation of other parts of the landscape.
- Plant edibles as part of your landscape, or even replace your lawn with a garden that can provide food for you and your family.
- Use manual equipment instead of gasoline-powered equipment for landscape maintenance. A standard walk-behind lawnmower puts out as much emissions as 11 cars, while a riding mower generates as much pollution as 34 cars. Even though mowers have small engines, they don't have emissions control systems like automobiles, although legislation is being developed at the state and federal levels that may change this requirement in the future.

1.3.6 Finding Better Energy Sources

After reducing your energy consumption as much as possible, investigate alternative sources of energy. These include distributed **renewable** energy systems such as wind turbines, hydropower, or **photovoltaics** you can install on site. Other options include purchasing green power if your utility provider offers it. This generally involves paying a small surcharge per kilowatt-hour to support the development of renewable energy resources such as large-scale wind, solar, and **geothermal** energy. Geothermal energy is derived from heat generated within the earth.

If green power is not available in your area (check www.green-e.org for availability), you can also reduce your carbon footprint through the purchase of carbon offsets. Carbon offsets can be purchased from certification companies and represent a certain amount of carbon that your purchase helps to reduce. For example, you may elect to purchase a carbon offset to mitigate the impacts of air travel, and you may even be asked if you'd like to buy an offset when you purchase a plane ticket. The money spent on offsets is used to help fund projects that reduce the amount of carbon emissions somewhere else. Carbon offsets can be used to plant trees, increase **energy efficiency**, or develop renewable energy sources. For example, a company may offset the carbon generated by its production processes by helping a school replace an old heating system. Spending money to reduce carbon emission somewhere else is one way to make yourself carbon neutral (that is, generating no greenhouse gases at all).

Reducing Your Carbon Footprint

Compare your carbon footprint with the list of actions in this section. Which actions can you take to reduce your carbon footprint? Which ones seem easiest to achieve? Be sure to consider what kinds of changes you will be most likely to sustain over time. Remember that changes in technology are often easier to sustain than changes in behavior.

DID YOU KNOW?

Making Greener Choices

Even small choices have an impact. The carbon footprint of the average cheeseburger is 6.6 pounds. This includes the energy required to grow the feed for the cattle, grow the produce, grow and process the grain, store and transport the components, and cook the burger.

101SA06.EPS

Review Questions

1. Reducing your overall energy use is a way to _____.
   a. save water
   b. save trees
   c. help the economy
   d. reduce your carbon footprint

2. A natural change in the green environment is _____.
   a. deforestation
   b. urban sprawl
   c. changes in the seasons
   d. aquifer depletion

3. One of the contributors to global warming is _____.
   a. aquifer depletion
   b. deforestation
   c. rising sea levels
   d. recycling

4. The increase in greenhouse gases is a result of _____.
   a. using biofuels
   b. burning fossil fuels
   c. using ethanol
   d. increased satellite use

5. The overuse of groundwater is leading to _____.
   a. loss of biodiversity
   b. global climate change
   c. aquifer depletion
   d. deeper coal mines

6. The water used to flush toilets is commonly _____.
   a. treated to the same quality standards as drinking water
   b. nonpotable
   c. treated with antibacterial chemicals
   d. treated with special salts

7. The percentage of waste generated by the average American household that could be recycled is _____ percent.
   a. 50
   b. 60
   c. 75
   d. 80

8. Products that perform in the top 25 percent of their product class in terms of energy use receive a(n) _____ rating.
   a. Green Energy
   b. Energy Star®
   c. Energy Efficient
   d. Green Performance

9. Airplanes are among the largest contributors to _____ emissions.
   a. helium
   b. nitrogen
   c. carbon
   d. oxygen

10. Spending money to help someone else reduce his or her emissions is called _____.
    a. carbon offsetting
    b. grid reduction
    c. reverse cycling
    d. impact control

2.0.0 ◆ BEST PRACTICES FOR CONSTRUCTION

Throughout recorded history, humans have constructed buildings to protect themselves and their possessions. Buildings provide shelter from adverse climate conditions such as rain, snow, wind, and temperature extremes. They also offer privacy and security. In addition to these roles, built facilities also serve the following purposes:

- Collection, treatment, and/or storage of solid, liquid, and gaseous waste
- Provision and distribution of pure water
- Processing and distribution of agricultural products into food
- Manufacturing and distribution of various products

The impacts of buildings on the environment have not always been obvious, but the effect over time is undeniable. Buildings are responsible for over 10 percent of the world's freshwater withdrawals. They also use 25 percent of the world's wood harvest and 40 percent of its material and energy flows.

In the U.S., 54 percent of all energy use is related to building construction and operation. Almost a third of all new and remodeled buildings suffer from poor **indoor air quality (IAQ)** due to emissions from various sources, **volatile organic compounds (VOCs)** released from building products, and **pathogens** spawned from inadequate moisture protection and ventilation. Poor indoor air results in variety of illnesses known collectively as **Sick Building Syndrome**, which is thought to cost more than $60 billion annually in lost productivity nationwide.

Nearly one-quarter of all ozone-depleting **chlorofluorocarbons (CFCs)** are emitted by building air conditioners. The processes used to manufacture building materials also contribute. Approximately half of the CFCs produced around the world are used in buildings. This includes refrigeration and air conditioning systems and fire extinguishing systems. CFCs are also in certain insulation materials. In addition, half of the world's fossil fuel consumption is attributed to the servicing of buildings. Lighting accounts for 20 to 25 percent of the electricity used in the U.S. annually. Offices in the U.S. spend 30 to 40 cents of every energy dollar for lighting. This makes it one of the most expensive and wasteful building features. Finally, the construction industry is responsible for 20 to 40 percent of the total municipal **solid waste** stream.

2.1.0 Facility Life Cycle

The life cycle of a built facility may range from 30 years to over 100 years. It typically runs through the following phases:

- *Planning or pre-design* – A facility's life cycle starts with an idea during the planning or pre-design phase. This phase focuses on defining the design and function of the building. The physical requirements of the building are determined, as is the budget and construction schedule. Also considered are any legal or regulatory constraints that should be taken into account during design. The outcome of the planning phase is typically a set of requirements. These describe the functional expectations the owner has for the facility.
- *Design* – The second phase of the facility life cycle is design. This is when the facility is transformed from an idea into a set of construction drawings and specifications. A design meeting called a **charrette** may be used to obtain input from project stakeholders.
- *Construction* – During construction, workers follow the set of construction documents to construct a building that meets the owner's requirements. The outcome of the construction phase is a completely functional building.
- *Operation and maintenance* – After construction, the operation and maintenance phase of the life cycle begins. The building is used to meet the design needs. This is typically the longest phase of the life cycle. Operation is the process by which the facility performs its intended functions. Maintenance consists of all actions performed on the facility to keep it in good operating condition. It includes activities such as changing light bulbs, servicing the building systems, and cleaning the facility. It also includes minor repairs or replacement of building components that break down.
- *Rehabilitation or end of life cycle and disposal* – At some point, a facility will no longer meet the requirements of its occupants. A possible choice is to rehabilitate or reconstruct the facility to improve its performance. This can be even more challenging than the original construction. Another possibility is to end the life cycle of the facility. **Deconstruction** and demolition are two options for ending the life cycle of a facility. Deconstruction is a planned, careful disassembly of the facility in order to sort waste and/or salvage building components for future use. Demolition is a more destructive process. Building components are removed from the site using various disposal methods.

Inventory the World Around You

Consider the built environment in which you're sitting right now. What features of the building do you think work well in terms of green performance? Which features do you think could be better? Working individually or in small teams, explore the building in which you are taking this class. Also, consider the site on which it is located. Begin by drawing a floor plan for the building. Add significant features of the site. Take an inventory of the following items and mark the location of each observation on your floor plan or site map. For each item, list why it's an example of excellence or how it could be improved.

- *Energy use* – Look around your building for technologies and features that save or waste energy. Pay special attention to the building envelope, which is the outside skin of the building including windows and doors. Examine the heating, ventilation, and cooling systems. Study the lighting and equipment in the building. What opportunities do you see to improve the energy performance of the building? What features already work well?
- *Water use* – Inspect your building for features that save or waste water. Pay special attention to faucets, fixtures, appliances, landscaping, and cooling systems that rely on water. What opportunities can you find to improve the water performance of your building? What things already work well?
- *Materials* – Now consider your building from the standpoint of materials. Examine the structure of the building, its enclosure, and interior and exterior finishes. Look at how the building is currently being used. What opportunities do you see to improve the performance of your building in terms of the materials used for construction or consumed for operations? What things are already satisfactory?
- *Indoor environment* – Look around the interior spaces of your classroom building. What about the spaces contributes to a comfortable and productive environment? What could be better? Pay special attention to lighting, acoustics, views, and indoor air quality.
- *Outdoor environment* – Step outside your building and inspect the site on which it is located. How could the building and its site be improved from the outside? In which ways is it already satisfactory? Pay special attention to the landscape (both the plants and the paved areas). Also consider the location of the building itself with regard to the people who use it. What transportation options exist now? What could be better?

When you've finished with your inventory, share your findings with the rest of the class. Have other students noticed things you didn't see?

Built facilities affect the green environment over their life cycles. *Figure 13* shows some of the relationships between a building and the environment during its life cycle. Note the import and export of materials, energy, and waste, all of which have impacts on the green environment.

You can change your actions to improve those impacts. Some impacts are due to the materials and energy used by the facility. Other impacts come from the waste streams that leave the facility. How the facility interacts with the site on which it is located creates impacts. How the facility interacts with its users also matters. The following **best management practices (BMPs)** represent a spectrum of options for improving the impacts of the built environment on the green environment:

- *Site and landscape* – This category involves choosing a good site for a facility. It also involves the specific placement of the facility on the site and the best design of landscaping to maximize energy savings and occupant comfort. Avoiding damage to the site is important. Restoring the quality of the site after construction is also important.
- *Water and wastewater* – This category addresses eliminating unneeded use of water. It also covers alternative water sources and wastewater sinks.

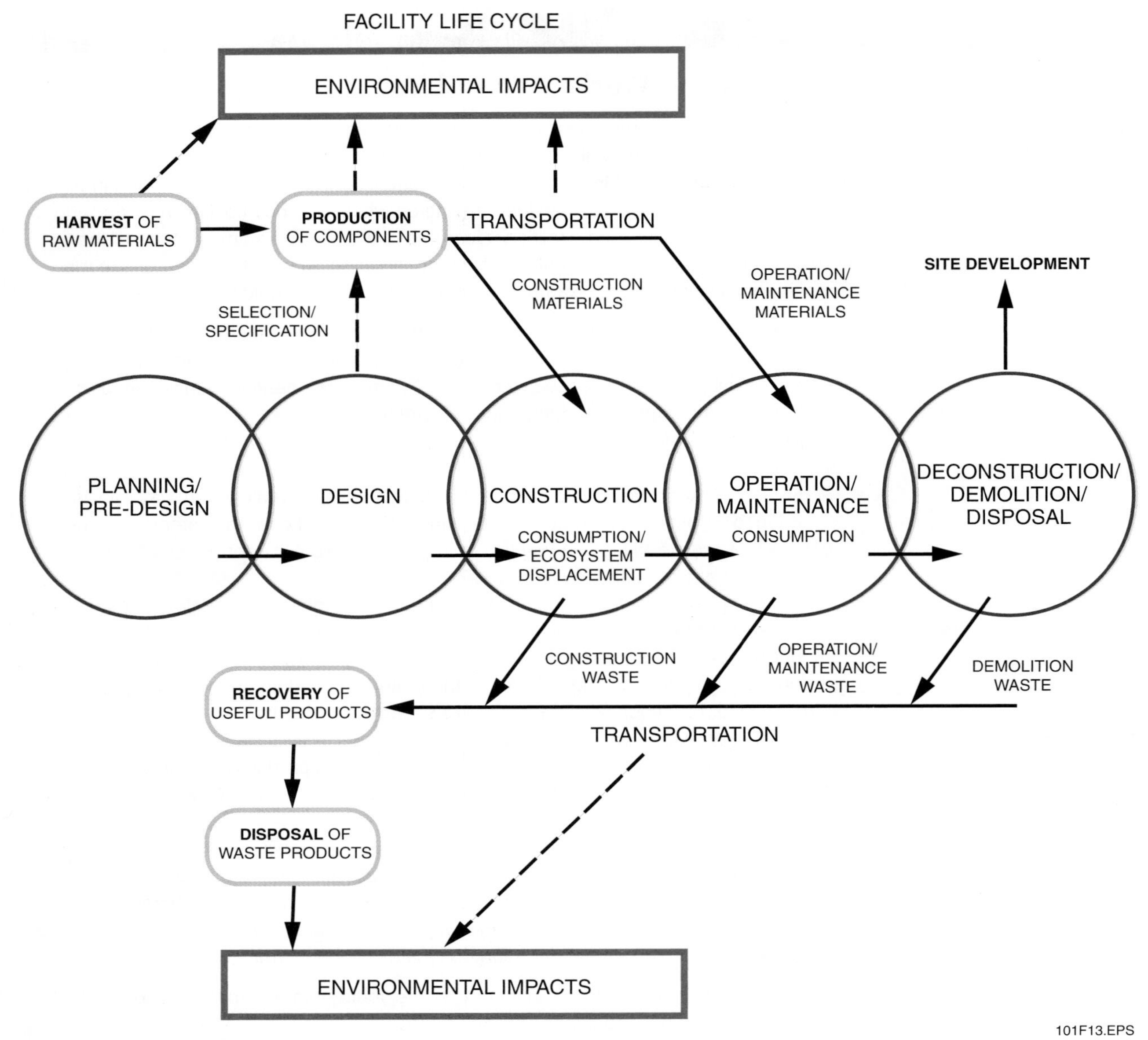

Figure 13 ◆ Interaction between a building and the green environment.

- *Energy* – This category addresses eliminating unnecessary uses of energy. It also covers methods of using energy more efficiently, balancing energy demands, and seeking alternative sources for energy.
- *Materials* – This category includes eliminating unnecessary uses of materials; using abundant, renewable, or multifunction materials; and seeking alternative sources of materials.
- *Waste* – This category includes eliminating or preventing waste, reusing waste within the facility system, sharing it with other systems, and storing it for future use.
- *Indoor environment* – This category includes preventing problems at the source, segregating polluters, taking advantage of natural forces, and giving users control over their environment.
- *Integrated strategies* – This category includes practices that result in multiple benefits from a single action. It includes capitalizing on construction means and methods. It also discusses making technologies do more than one thing and exploiting relationships between systems.

Each of these categories is shown in *Figure 14* and described in detail in the following sections.

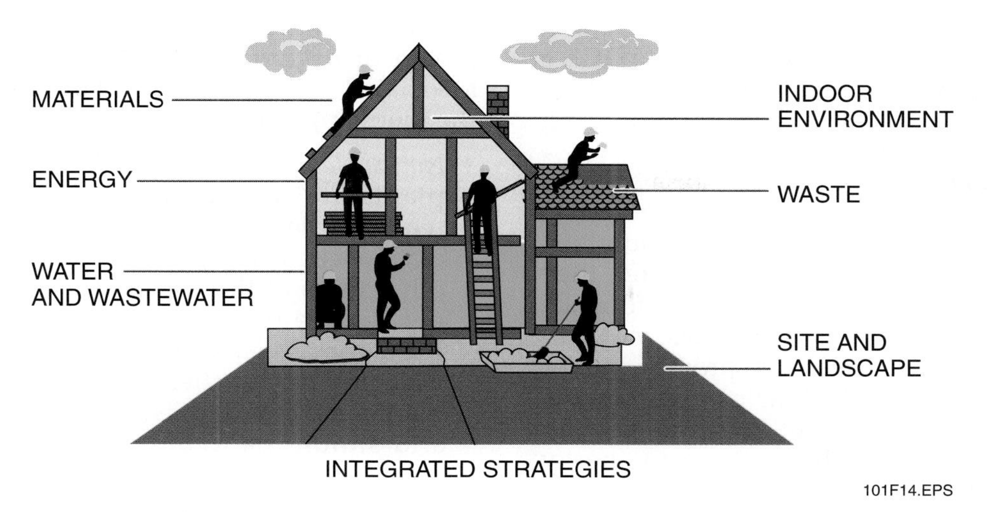

Figure 14 ◆ Categories of best practices.

2.2.0 Site and Landscape Best Practices

The first category of best practices deals with the facility site and landscape. The most important decision for a building is the selection of a site. After that, the next most important decision is where to put the building on the site. These two decisions affect the building's performance over its life cycle. Next, the development of the site must be considered. It should use low-impact principles and features. Finally, it should ensure that ecosystems on the site are restored to their best quality. This will help to keep the site functioning well.

2.2.1 Site Selection

The choice of site affects the building's energy use and environmental impact. It also governs travel to and from the facility. For example, choosing a site in an already developed area provides building occupants with access to existing stores and services. It may even allow them to take advantage of walking, bicycling, or transit to access the building instead of having to drive a car. It can also provide a positive contribution to an existing neighborhood. Choosing a site that is a **brownfield** is a possibility. A brownfield is a site that may have real or perceived environmental contamination. Cleaning up a brownfield represents a positive step for the community. It may also provide a tax credit or development incentive for the builder.

Avoid sites with valuable ecological resources or higher levels of risk from environmental damage, such as sites with **wetlands** or habitats of threatened or **endangered species**. This preserves these resources and prevents the need to mitigate or help correct any impacts. It also saves considerable expense. Choose sites that are well out of flood plains or areas where mudslides or wildfires are common. Finally, consider the long-term risks of resource depletion. For example, the water supply in areas such as the southwest U.S. and even the southeast, such as Atlanta, is becoming scarce. This may mean restrictions on how facilities are developed and used in the future.

Some sites can share common resources with other sites if the facilities have different hours of operation. For example, a bank that is only open during the day may allow parking for a nightclub or restaurant at night. This eliminates the need for two separate parking lots. Choosing sites in already developed areas provides access to the existing infrastructure, which includes streets as well as water, power, and sewer lines. This can save time and money and reduce the negative effects on the site.

2.2.2 Building Orientation

A building's relationship to the sun has a large impact on energy use. In fact, it can save 30 to 40 percent in heating and cooling costs over the life cycle of the building. *Figure 15* shows a building in relation to the sun's path in the Northern Hemisphere. The building is oriented with its long axis running from east to west. Most of the windows are on the south side of the building. In the summer, the sun passes higher in the sky and the overhangs on the building help shade the windows, which keeps the building cooler. In the winter, the sun passes lower in the sky. This allows sunlight to enter the windows under the overhangs and provides free heat. This is an example of passive solar design. Passive solar design takes advantage of the sun's energy to provide heat without using electrical or combustion energy.

Trees can also be used to provide free cooling. Deciduous trees on the south, east, and west sides of a building can provide shade in the summer, which reduces cooling costs. In the winter, they lose their leaves and allow solar energy to reach the building. This energy provides warmth and heat gain in the winter. Trees also provide shelter from prevailing winds, which helps to reduce the heating load on the building. Preserve existing trees whenever possible. In addition to energy benefits, trees also have positive effects on water retention, air quality, and property values.

Another important consideration when selecting a site is avoiding areas that are difficult to develop. For example, areas with steep slopes are more difficult to develop than flat areas, and require more earthwork and/or larger foundations. The same is true for areas with high water tables. After examination of the entire site, thought must be given to both initial and long-term costs of the building location.

2.2.3 Landscaping

Landscaping is divided into two main categories, the softscape and the hardscape. Softscape refers to the vegetated parts of a site, such as grass, trees, and bushes. Hardscape refers to areas that are paved or otherwise developed. Best practices associated with low-impact landscaping include the following:

- *Native plants* – Use native plant species. Native plants are well adapted to the climate and require little or no irrigation, fertilizer, or pesticides once established. Avoid exotic plants from other parts of the world. They often out-compete native plants and can become invasive. Kudzu is a well-known invasive species common in the southeast U.S. Many common plants, such as English ivy, are also invasive and should be avoided. Visit www.plants.usda.gov and click on the links for noxious and invasive plants. This website also has information on native plants that may do well in your area.
- *Zoned landscaping* – Group plants with similar needs. This is called zoning. Zoned landscaping allows you to focus irrigation, fertilizer, and pesticides only on the plants that require it. It can also help reduce the risk of wildfire around buildings. This is accomplished by keeping areas near buildings clear of flammable vegetation and debris.
- *Mulching* – Use bark chips, pine straw, or other natural materials to suppress weeds and retain moisture. This reduces the need for irrigation and pesticides. It also helps stabilize new plants until they become established.

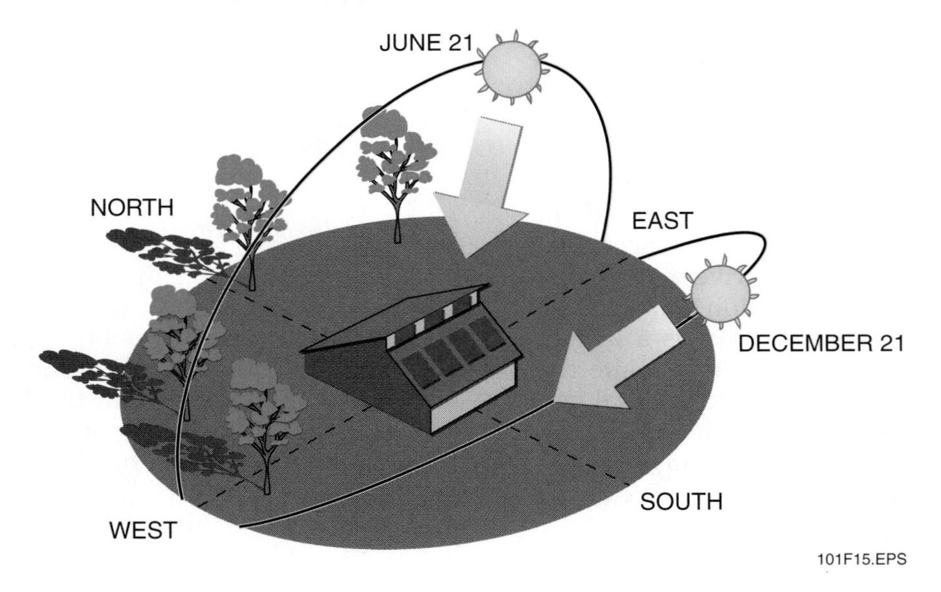

Figure 15 ◆ Building oriented to take advantage of the sun's path.

- *Xeriscaping* – **Xeriscaping** is the selection of native plants that require no irrigation once they are established. It is common in the southwest, where xeriscape yards of rock, sand, and native plants replace green lawns that would require high volumes of water in order to survive.
- *Runoff control* – Use vegetated areas to absorb and treat stormwater runoff from paved areas. **Bioswales** and rain gardens are engineered basins that collect stormwater. They are similar in purpose except that rain gardens tend to have a more formal layout of plants and stones, while bioswales are planted using a more natural approach. Once collected, the water is percolated slowly back into the soil (*Figure 16*). Plants within the bioswale also help to remove contaminants from runoff. These pollutants would otherwise have to be treated at a wastewater treatment plant to avoid contaminating local water streams.
- *Permeable pavement* – Permeable pavement captures stormwater and allows it to percolate back into the soil (*Figure 17*). Options include **pervious concrete** (concrete that contains spaces through which water can move), stabilized soil, and grid systems. Stormwater that drains over pavement may become contaminated and require collection and treatment. This is particularly true in areas where vehicles drive or park. These contaminants include engine leaks, rubber and asbestos particles, and other residues. When concentrated in runoff, these contaminants can pollute local streams and rivers. Stormwater runoff can also cause stream damage due to an increase in water temperature when it contacts pavement. This leads to thermal pollution of local streams, which can harm aquatic plants and animals. Some permeable pavements provide basic treatment of contaminants. This is done through plant roots or microorganisms in the pavement pores.
- *Light-colored pavement* – Pavement that is light in color helps minimize the amount of heat absorbed from the sun. Light colors have high **albedo** values. The term albedo describes the extent to which an object reflects light from the sun. It is a ratio with values from 0 to 1. A value of 0 is dark (low albedo), while a value of 1 is light (high albedo). High-albedo pavements help to minimize the urban heat island effect, which is caused in part by dark asphalt and buildings that absorb rather than reflect heat. Using light-colored concrete can reduce overall temperatures during summer months and lower the air conditioning requirements of surrounding buildings. *Figure 18* shows high- and low-albedo pavements.
- *Alternative transportation* – Some landscaping features promote the use of alternative transportation (*Figure 19*). These include sheltered transit stops, bike racks, and bike paths and sidewalks or pedestrian trails. Setting aside special parking for carpools or **alternative fuel** vehicles rewards drivers for making greener choices. One type of alternative fuel is **biofuel**, which is commonly made from plants.

101F16.EPS

Figure 16 ◆ A bioswale receives stormwater runoff from a parking lot.

101F17.EPS

Figure 17 ◆ Permeable paving options include pervious concrete and grid systems that allow water to pass into the soil.

Figure 18 ◆ High- and low-albedo pavements.

Figure 19 ◆ Amenities to promote alternative transportation.

2.2.4 Restoring Ecosystems

After developing a site using green amenities, restore natural ecosystems to the extent possible. Set aside areas of the site to remain undeveloped. Native species require little or no maintenance. They provide landscape beauty and serve as a natural habitat for various creatures, including birds and butterflies. Whenever possible, work with adjoining sites to create common undeveloped areas, including local streams and waterways.

2.3.0 Water and Wastewater Best Practices

The second category of best practices involves the sources and uses of water for a facility. These practices also deal with sinks for a building's wastewater. Nearly all buildings require a source of water to meet the needs of occupants. This includes water for drinking, washing, and waste disposal.

Water may appear to be abundant, but less than 3 percent of all water on Earth is fresh water. The rest is salt water, which is mostly unusable. Of the small amount of fresh water available, 69 percent is trapped as ice and snow cover and 30 percent is stored as groundwater. Less than 1 percent is available on the surface as freshwater lakes and rivers. The **hydrologic cycle** continuously replenishes the freshwater supply. This process has five phases: condensation, infiltration, runoff, evaporation, and precipitation. In many areas, the rate of use exceeds the recharge rate, leading to aquifer depletion.

To conserve this precious resource, the most important action you can take is to eliminate unnecessary uses of water. This can be followed by increasing the efficiency of the water you do use. After determining how to reduce the need for water as much as possible, the next steps are to look for alternative sources of water and seek alternative ways to remove and treat wastewater.

2.3.1 Reducing Water Use

If you are involved with an existing building, complete a water audit. This will help find leaks and identify opportunities for improvement. Your local water authority may be able to provide information on completing a water audit.

Another way to eliminate water use is to install waterless toilets and urinals. Waterless urinals (*Figure 20*) take advantage of the fact that urine is a liquid and does not require water to be conveyed through wastewater pipes. Waterless urinals use a special trap that allows urine to pass through, but prevents sewer gases from escaping.

Waterless toilets are also available. For example, composting and incinerating toilets both function without the use of water. Composting toilets convert human waste into harmless compost that can be used to improve the soil. Incinerating toilets use electrical energy or other fuel to convert the waste into ash. Note that incinerating toilets use a considerable amount of energy, which should be taken into account as a tradeoff against the water savings.

2.3.2 Increasing the Efficiency of Water Use

Many types of fixtures are available to reduce water use in showers and sinks. These include low-flow or aerated fixtures and alternate types of controls. Aeration is the introduction of air into the water flow. It is a common technique to reduce the actual flow of water in faucets and showerheads while maintaining the perception of high-volume flow. It works well for hand washing and showers, but can be frustrating when filling a pot with water for cooking or cleaning. Some aerated kitchen faucets allow the user to control the degree of aeration and restore full flow when needed.

101F20.EPS

Figure 20 ◆ Waterless urinal.

Other types of alternative controls include foot-operated sinks. These allow precise control of flow timing while leaving hands free. More precise automated flow controls are also becoming available for sinks and toilets. The EPA's WaterSense® program provides a labeling system to help identify low-flow fixtures and controls (see www.epa.gov/watersense).

Toilet technology has improved considerably over the last decade. There are now multiple options available to improve the efficiency of water use. In addition to conventional low-flow toilets, several manufacturers now offer dual-flush toilets. These allow a full-volume flush for solid waste and a half-volume flush for liquids. Automated dual-flush control units are also available. These units dispense either a full-volume or low-volume flush depending on how long the user is within range of the flush sensor. Manual controls are also included to allow users to control the flush level if desired.

When selecting water sources for landscaping, consider drip irrigation or irrigation controlled by moisture sensors and timers. Drip irrigation provides a slow release of water directly to the root zone of the plant. This results in less evaporation than a typical sprinkler system. Moisture sensors and timers control the operating periods of irrigation systems. This ensures that water is only dispensed when needed. It also allows watering during the early morning when water has time to reach plant roots before evaporating. With either system, be sure to properly locate and maintain all components of the system to avoid waste.

Water-efficient appliances, such as high-efficiency dishwashers and washing machines, reduce water use by up to 50 percent over traditional units. Since these units use less hot water, they also save energy and are reviewed under the EPA's Energy Star® program. See www.energystar.gov for a list of washing machines and dishwashers that meet Energy Star® water-saving criteria.

2.3.3 Finding Alternative Sources of Water

A third option for greening your use of water is to seek alternative sources of water. Focus on uses that do not require water to be treated to drinking water standards. Much of the water used in homes and businesses does not require this high level of water quality. Think about the extra energy and treatment required to use drinking water to flush a toilet.

One alternative source of fresh water is known as **rainwater harvesting**. It involves capturing and using rainwater. Rainwater harvesting has been used on a small scale since development began. Larger scale systems have been used in arid parts of the country for decades and are likely to become more common as water scarcity increases. *Figure 21* shows a rainwater harvesting system. These systems require a collection surface, typically a roof that is relatively debris-free. The surface must also be made from a chemically stable substance. Older metal roofs should not be used as they may have joints sealed with lead-based solder, which could contaminate the water. Asphalt shingles are also not recommended since they tend to shed particles as they age. Slate, synthetic, and lead-free metal roofs are all good collection surfaces.

In addition to the collection surface, rainwater systems require components to provide filtration, storage, and overflow. Most existing roof drainage systems can be retrofitted for rainwater harvesting. Depending on rainfall levels, a roof can capture tens of thousands of gallons of water per year.

Another alternative source of water is **graywater**. This is used water from sinks, showers, and laundry facilities. It is called graywater because it contains a negligible amount of human pathogens. Full-building graywater systems collect and filter water for irrigation or toilet flushing (*Figure 22*). Graywater systems must be carefully designed to meet code requirements. This is because graywater is an excellent harbor for bacterial growth while in storage. Separate piping must be used and provisions made to balance supply with demand. Graywater that remains in the system after a specified period must be disposed of rather than reused.

Small-scale graywater systems are also available. They capture water from one fixture, such as a shower, and divert it for use in another fixture, such as a toilet. Some systems fit under a bathroom sink and work well for retrofits.

Water from toilets and dishwashers is considered **blackwater**. This is because it contains pathogens. The pathogens are from either human waste or from animal fats associated with dishwashing. Blackwater requires separate piping and additional levels of treatment before reuse.

2.3.4 Finding Alternative Sinks for Wastewater

The final opportunity to green a building's water systems involves finding other uses or alternative treatments for wastewater. A building's wastewater stream also includes stormwater. Stormwater may be contaminated by materials from parking lots and roofs. It is also likely to be much hotter than local water streams. Alternative systems are available to treat this type of wastewater.

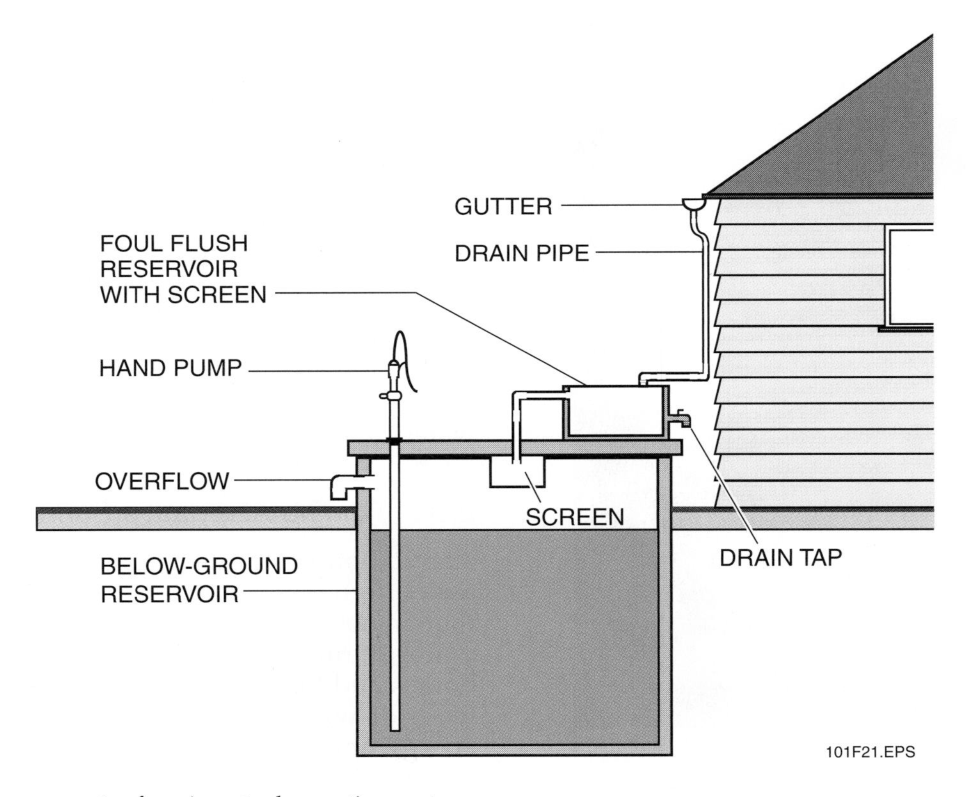

Figure 21 ◆ Components of a rainwater harvesting system.

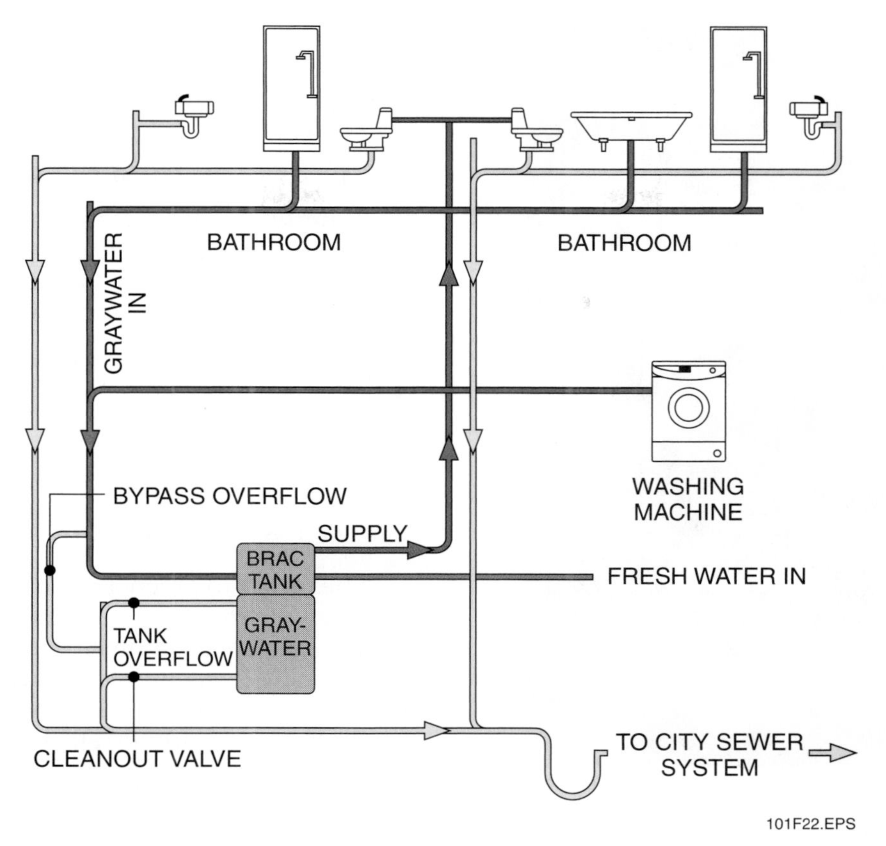

Figure 22 ◆ Components of a building-scale graywater system.

Graywater systems can capture water that has not been heavily contaminated. This water can be redirected for certain uses without treatment if allowed by local codes. These systems can also be the first step in a more comprehensive system. Lightly contaminated water is recycled for heavier use, and then treated on site using a separate system.

One type of alternative wastewater technology is a plant-based constructed wetland. Constructed wetlands have been used in a variety of climates and applications. *Figure 23* shows a constructed wetland for a small town with about 900 residents. This series of ponds is filled with plants that provide increasing levels of water treatment. A primary treatment step removes solids from the water stream before it enters the ponds. After treatment, the clean water is discharged to the local water stream.

Smaller plant-based or bio-based wastewater treatment systems can be used to treat the wastewater from a single building. One example of a bio-based system is called a Living Machine®. Living Machines® combine plants, bacteria, and other organisms into a simulated ecosystem that uses wastewater as a nutrient source. *Figure 24* shows a bio-based wastewater treatment system at the Rocky Mountain Institute in Snowmass,

Figure 23 ◆ Constructed wetland wastewater treatment system.

101F24.EPS

Figure 24 ◆ Bio-based wastewater treatment system.

Colorado. This system includes hedgehogs and lizards as part of the indoor ecosystem, and it even produces bananas in winter.

2.4.0 Energy Best Practices

The energy used in homes and buildings has considerable impacts on the green environment. A variety of energy best practices can be used to reduce these impacts, including:

- Avoiding unnecessary energy use
- Increasing the efficiency of energy use
- Balancing electrical loads
- Seeking alternative energy sources

2.4.1 Avoiding Unneeded Energy Use

The best way to green a building's energy use is to reduce or eliminate the demand for energy. The oldest approach involves setting back your thermostat and wearing a sweater in the winter or opening windows to take advantage of the breeze in the summer. Other strategies include timers and occupancy sensors that can be used to control the lighting or temperature when a space is unoccupied.

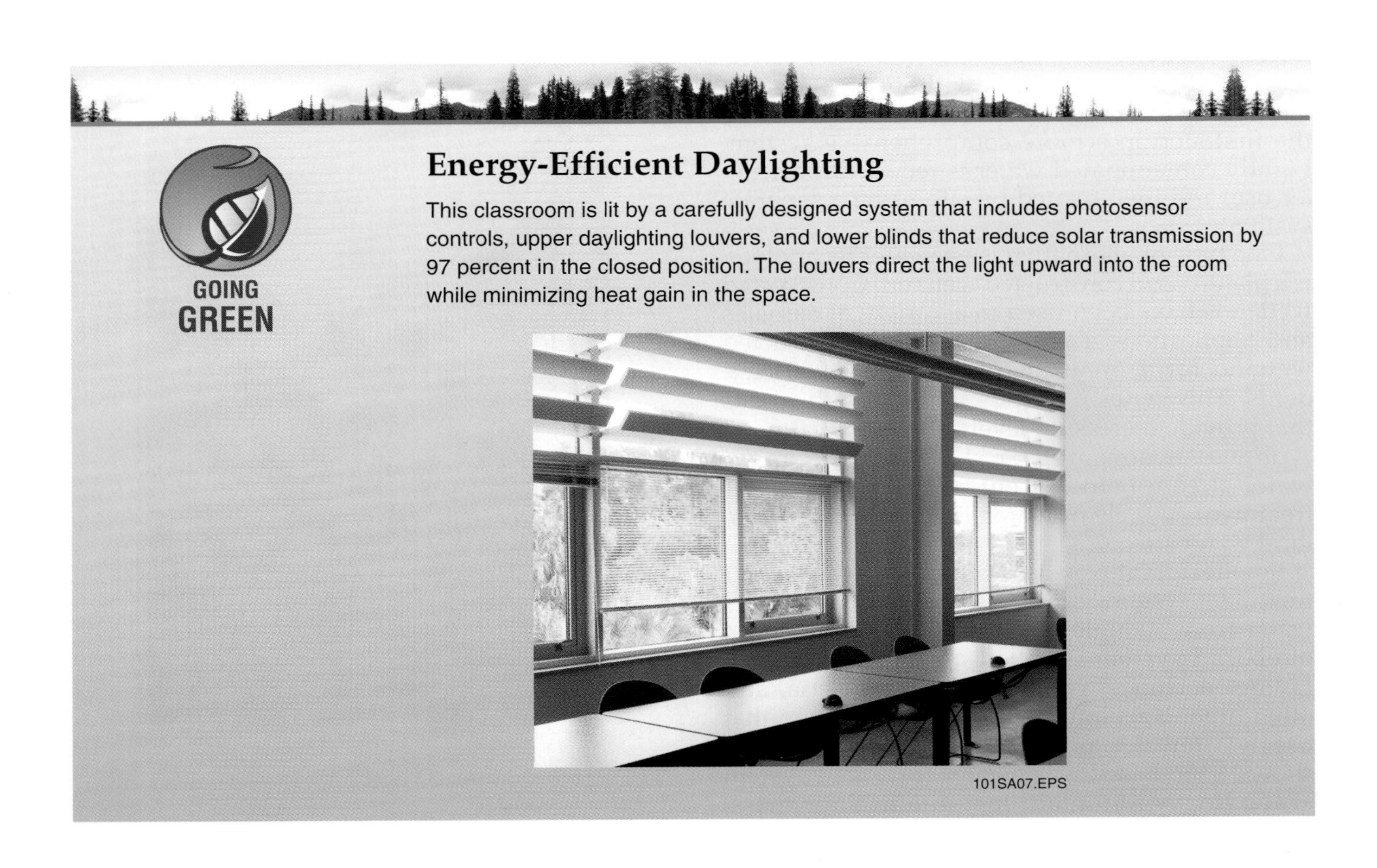

Energy-Efficient Daylighting

This classroom is lit by a carefully designed system that includes photosensor controls, upper daylighting louvers, and lower blinds that reduce solar transmission by 97 percent in the closed position. The louvers direct the light upward into the room while minimizing heat gain in the space.

101SA07.EPS

Other strategies reduce demand with no changes required in occupant behavior. These include passive heating or cooling, daylighting, and heat recovery ventilators (HRVs). HRVs capture waste heat from exhaust air and use it to preheat incoming air. Some systems also equalize humidity in a similar way. Other systems take advantage of the **thermal mass** of building materials. Thermal mass is a property that allows a material to absorb heat and then release it slowly. For example, adobe houses have thick walls that absorb heat during the day and release it at night.

Another simple strategy is a high-albedo roof (*Figure 25*). These roof materials and coatings reduce heat gain by reflecting solar energy rather than absorbing it. As *Figure 25B* shows, these surfaces need not be white and shiny. New coatings are available that reduce heat gain even with darker color roofs. In many cases, the albedo value of a roof can be increased at no cost. This is done by simply choosing a different color for roof finishes during design.

High-performance building envelopes can also eliminate unnecessary uses of energy for heating and cooling. A building's envelope is simply the outer shell that protects it from the elements. If you added up the area of all cracks and openings in the building envelope of a typical American house, it would equal about one square yard. That is the equivalent of leaving a window fully open year-round, even when the heater or air conditioner is on. Ways to improve a building's envelope include the following:

- Increasing insulation in wall cavities and attics
- Sealing cracks and using **vapor-resistant** barriers to reduce infiltration
- Installing high-efficiency windows and doors
- Using low-emissivity paint or radiant barriers in attics to reduce heat gain

All of these methods can reduce the demands on facility space conditioning equipment. They also increase occupant comfort.

Finally, using good control systems for lighting can help to eliminate the use of energy during hours when daylight provides enough light. Dual switching involves wiring light switches so that combinations of lamps can be switched on or off depending on how much light is needed. This helps users to adjust the light to appropriate levels. Properly labeling the switches allows for ease of use.

A. HIGH-ALBEDO MEMBRANE ROOF

B. HIGH-ALBEDO METAL ROOF

101F25.EPS

Figure 25 ◆ High-albedo roofs.

2.4.2 Increasing the Efficiency of Energy Use

There are many opportunities for increasing the efficiency of energy use. One way is proper sizing of heating, ventilation, and air conditioning (HVAC) equipment. Oversized equipment operates at peak efficiency only a few days per year. A better choice is the use of two parallel units. Under normal conditions, only one unit is required and it operates at peak capacity. Both units are used to meet peak demands. The initial cost of two systems is offset by the energy savings.

An often-overlooked method of increasing efficiency is proper operation and maintenance. Even simple maintenance, such as cleaning the lenses on lighting fixtures, can greatly increase light output and reduce the need for supplemental lighting.

Processes such as continuous **commissioning** can be used for existing buildings. Commissioning is a procedure for system setup that examines each component to ensure it is operating properly. It can be used to identify ways to improve efficiency with existing equipment. Continuous commissioning typically pays for itself in less than two years. Operator training is essential to ensure that building equipment is used most effectively.

Often, energy audits identify opportunities to retrofit existing systems with more efficient ones. A common retrofit with rapid payback in many buildings is a lighting retrofit (*Figure 26*). You can begin by replacing older magnetic ballasts with electronic ballasts in fluorescent lights. Another way is to convert existing fixtures from T-12 lamps to T-8. Other lighting retrofits may include installing solid state LED lighting fixtures or T-5 fluorescent lighting. These retrofits require replacement not just of lamps but also of fixtures themselves.

Replacing tank-type hot water heaters with tankless models is another way to save energy. Tankless models cost more initially, but typically last longer than tank-type models if properly maintained. They also eliminate standing heat loss since water is only heated as needed.

High-performance HVAC systems also save energy. Variable speed fans, pumps, and motors enable these systems to operate at optimum levels. This makes the system cycle on and off less frequently, which increases user comfort and saves energy. Optimized distribution systems also help reduce heating, cooling, and ventilation costs. Sealing and insulating ducts is an important way to ensure that the energy used to condition air actually benefits the users of the building.

101F26.EPS

Figure 26 ◆ Lighting retrofits are an easy way to save energy.

GOING GREEN

Reducing the Cost of the Brooklyn Bridge

The Brooklyn Bridge will soon go green with LED lights, part of New York City's plan to save energy and cut greenhouse gases. The 160 LED fixtures will each use only 24 watts of power instead of the 100-watt mercury bulbs in place now. This will save on energy costs and reduce carbon dioxide emissions by about 24 tons each year.

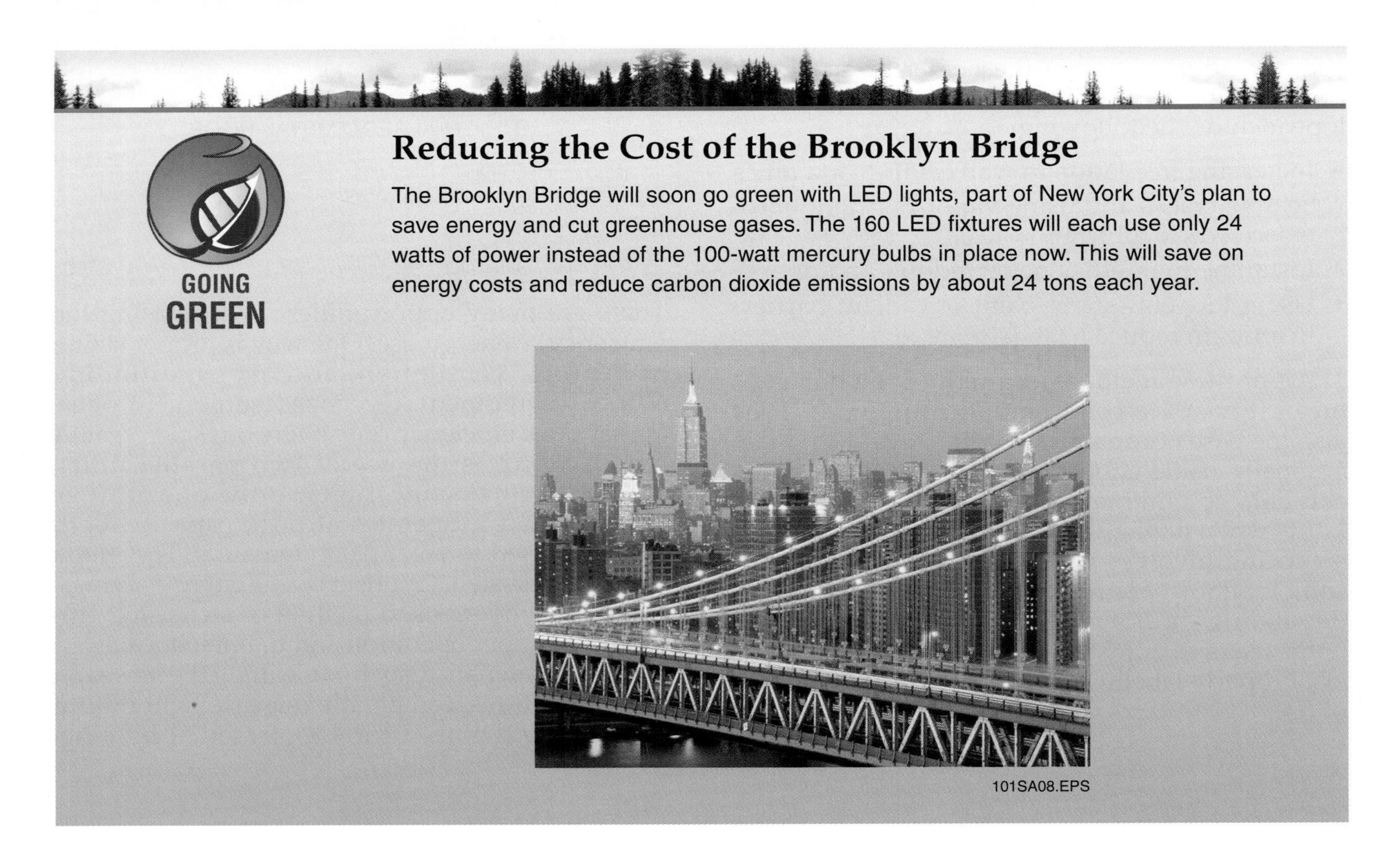

101SA08.EPS

Finally, ultra high-efficiency appliances and equipment can quickly pay for themselves in energy savings. Look for Energy Star® appliances when making replacements or buying new equipment.

2.4.3 Balancing Electrical Loads

The electrical grid is a complex system of power plants and distribution networks. The end systems that consume the power are also complex. To keep the system functioning, energy supply must remain balanced with energy demand. If this does not happen, the system becomes unstable. When demand exceeds supply, additional generators are brought online to meet the additional need. If these peak generators cannot meet demand, the system may experience brownouts, which are noticeable losses in system voltage that may damage equipment such as motors. In the worst case, the system becomes so unstable that safety mechanisms activate to take parts of the grid offline. This is known as a blackout.

Load balancing is a way to limit the loads placed on the power grid. This allows peak power generators to be smaller. It also reduces the need for new power plants. The highest demand for energy is typically during the middle of the day in the hottest part of summer. At this time, industry is operating at maximum output and commercial buildings are operating air conditioning at maximum capacity. Shutting down unnecessary equipment during these periods is one way to reduce the need to bring peak generators online.

Peak shaving reduces demand during peak times of the day and shifts it to off-peak times, such as at night. Utilities often charge lower rates during off-peak times. This provides an incentive for users to shift demands to these periods. Certain kinds of thermal storage systems are designed to take advantage of peak shaving. Energy is used at night to super-cool a thermal storage medium. This medium then absorbs heat to provide cooling during the day.

Energy management systems use electronic controls to balance loads during peak periods. They do this by reducing the amount of power sent to equipment that is tolerant of voltage variations, including certain types of air conditioners, pumps, fans, and motors. They also maintain a steady stream of power to equipment that requires a constant voltage input, such as computers. Depending on the local or regional climate for power production, the price for electricity during peak hours can be quite high. In many areas, an investment in energy management equipment can

Statue of Liberty Goes Green

The Statue of Liberty celebrated her 120th birthday in 2006 by going green. All of the electricity that powers the Statue of Liberty is now green power, provided by windmills in West Virginia and Pennsylvania.

101SA09.EPS

rapidly pay for itself in terms of reduced energy costs. Often, utility companies will finance or invest in these systems for built facilities. They may also provide technical assistance to support implementation.

Finally, information can be a useful way to balance loads and shave peak demand. Large institutions such as universities often pay for power based on time-of-day usage. During the hottest parts of the year, rates can easily triple when demand for power is highest. Some institutions send out email reminders to shut down unnecessary equipment and lighting during these periods (*Figure 27*). This saves both electrical energy and money by reducing use during high-price periods.

WE NEED YOUR HELP!

Energy Advisory

Electricity Price Caution is in effect today for the time period from 3 pm to 7 pm.
Electricity Price Caution is issued when the electricity prices are 3 to 6 times higher than normal.

Here are some actions you can take to help reduce our electricity consumption:

1. Activate the energy saving or "sleep" mode on computers and copiers.
2. Turn off your computer monitor when you are away from your desk for more than 15 minutes.
3. Turn off lights when out of your office or cubicle.
4. Turn off lights in unused common areas such as copy rooms, break rooms, conference rooms, unoccupied rooms, and restrooms.
5. If you have control over the thermostat setting for the air conditioner, raise it by two degrees during the peak hours. Consider raising the level of the thermostat further when your facilities are unoccupied.
6. Shut off nonessential machinery, computers, and other equipment.
7. Consider reducing the number of copiers available for use during peak hours.

XYZ Power Company – Prices for Today

Hours	cents/kWh
1:00	2.1203
2:00	2.0594
3:00	2.0235
4:00	2.0263
5:00	2.0278
6:00	2.0213
7:00	2.1258
8:00	2.1257
9:00	2.1637
10:00	2.2745
11:00	2.4378
12:00	3.3367
13:00	4.9845
14:00	8.0461
15:00	11.3748
16:00	12.5938
17:00	12.4085
18:00	11.2554
19:00	8.9338
20:00	7.2654
21:00	6.3741
22:00	4.9265
23:00	3.3319
24:00	2.2695

Average price = 6.0211 per hour at end interval

101F27.EPS

Figure 27 ◆ Using information to encourage energy conservation for peak shaving.

2.4.4 Finding Alternative Energy Sources

After reducing demand, optimizing efficiency, and balancing loads, the last step is to explore alternative sources of energy. There are two primary ways to obtain alternative energy: buying energy from a green power provider or generating power on site. Green power providers generate electricity from renewable energy sources such as wind power and solar power. They avoid nonrenewable power sources such as fossil fuels. To see if certified green power is available in your area, visit the Green-e website at www.green-e.org. Green-e is a nonprofit organization that certifies the renewable energy power companies sell to consumers. *Figure 28* shows the Green-e Energy Certified logo that companies can use on the certified renewable energy products they sell.

The second option is to explore the possibility of on-site renewable energy generation. This includes photovoltaics, wind turbines, gas-fired microturbines, and fuel cells. *Figure 29* shows a roof-mounted wind turbine providing power to a commercial building.

Stand-alone power generation systems require costly and high-maintenance battery storage for excess power. Given the difficulty of storing electrical energy, the most effective way to use on-site alternative power systems is with a grid intertie, which is a connection to the local utility grid.

101F28.EPS

Figure 28 ◆ The Green-e Energy Certified power provider logo.

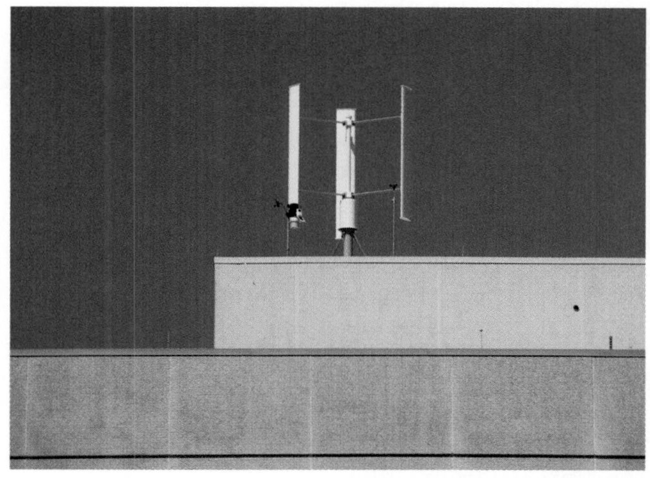

101F29.EPS

Figure 29 ◆ Small-scale wind turbine on a commercial building.

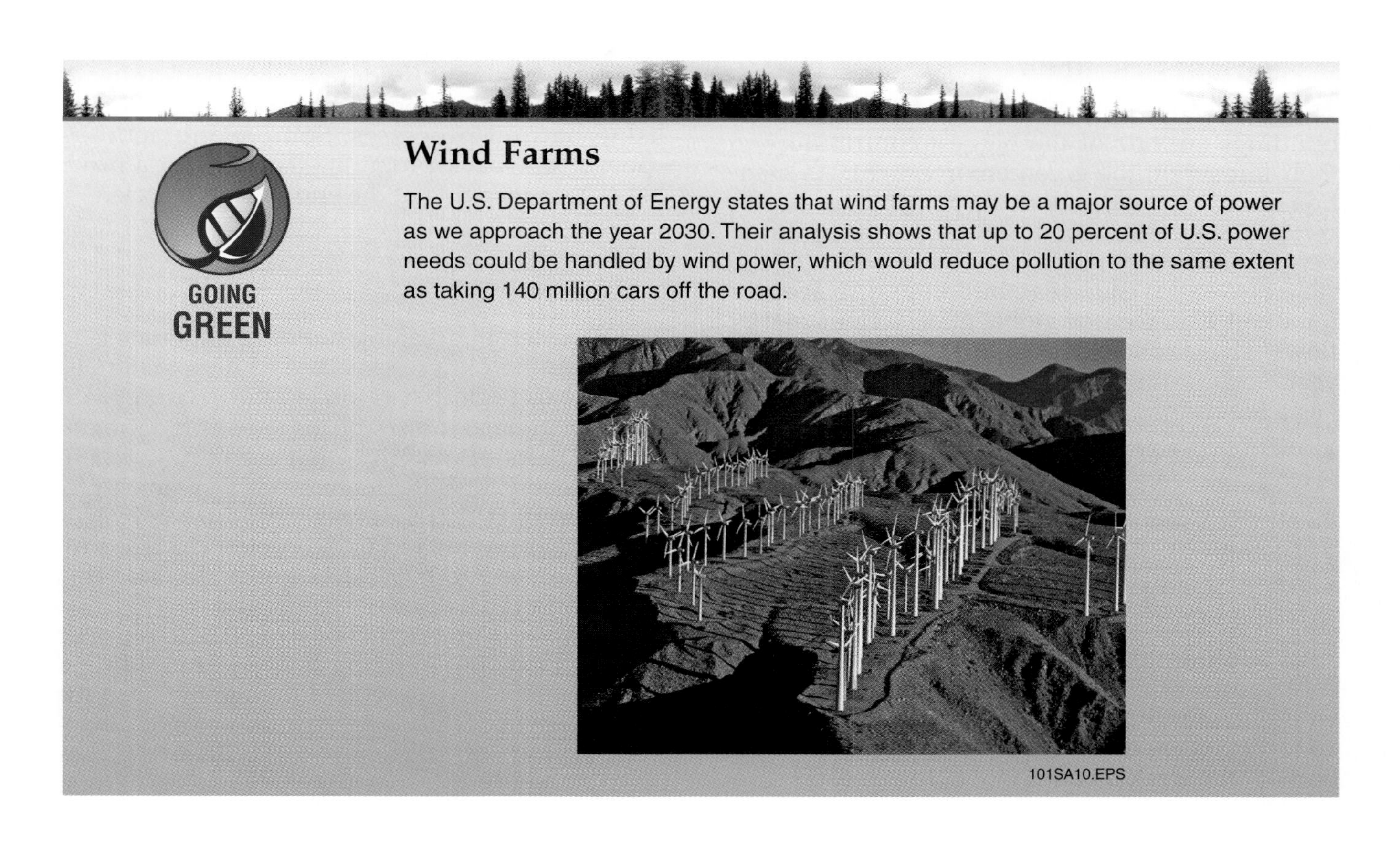

Wind Farms

The U.S. Department of Energy states that wind farms may be a major source of power as we approach the year 2030. Their analysis shows that up to 20 percent of U.S. power needs could be handled by wind power, which would reduce pollution to the same extent as taking 140 million cars off the road.

101SA10.EPS

It allows you to route excess power back to the grid when you generate more than you can use. You can also continue to buy power from the utility when your on-site system does not meet your demands. With photovoltaic systems in particular, peak capacity falls at the same time of day as peak demand. This happens in the afternoon. In some areas, you may even be able to sell your power. If you live in a part of the country where there is not much excess generating capacity, utilities may provide financing and technical assistance to help pay for renewable energy systems. Check online at www.dsireusa.org to see what programs are available in your area.

With demand for energy only likely to grow in the future, you need to be aware of opportunities to manage consumption of electrical power in built facilities. Energy investments save money over the life cycle of a facility. They can also increase the ability of the facility to withstand fluctuations in power supply. This reduces vulnerability to natural disasters and other threats. As dependence on power grows, the vulnerability of power systems will increase as well. Facilities with well-managed, minimal energy requirements will be the least vulnerable to fluctuations in power (and prices) from utility suppliers. These facilities will use on-site power generation as well as passive systems for lighting, heating, and cooling.

2.5.0 Materials and Waste Best Practices

The materials used to construct and maintain buildings are one of the biggest contributors to their impact on the green environment. As mentioned earlier, buildings are responsible for 25 percent of the world's wood harvest. Current rates of deforestation amount to an area the size of the state of Georgia each year. Buildings also account for about 40 percent of global material and energy flows. That adds up to over three billion tons each year. Each product or material used in a building has a life history that includes the following:

- Harvesting of all the raw materials required to make it
- All the **byproducts** generated during its production
- All the energy used to transport its components from place to place

In addition to raw materials, solid waste also contributes to a building's impact on the green environment. Buildings are responsible for up to 20 to 40 percent of the solid waste stream in some parts of the United States.

Greening your resource use starts by eliminating the unnecessary use of new materials. This includes using materials that come from waste, **salvaged**, or recycled sources. It also includes reusing and adapting existing buildings instead of building new ones.

The second step is to use materials more efficiently. Construction methods that reduce waste contribute to this goal. A whole new generation of **multi-function materials** is now available that serve as structure, enclosure, insulation, and other functions. Using these products reduces the amount of raw material needed to meet building needs. They also save on shipping and packaging. These materials are pre-engineered to minimize waste on site.

Seeking better sources for the materials we use can help ensure an ongoing supply of products to meet future needs. Switching to abundant, renewable materials helps to preserve the limited supply of nonrenewable resources. Ensuring that materials are **sustainably harvested** means that the supply of those materials can continue indefinitely. Using **local materials** helps to reduce the impacts of transportation of raw materials and supports local economies.

Finally, finding better sinks (destinations) for waste materials generated by the building helps the green environment. Some of the ways to improve performance in this area include on- and off-site recycling, **biodegradable** or **reusable** packaging, and salvaging or deconstructing building systems. Biodegradable packaging is made of natural materials that readily break down using natural processes. Some manufacturers sponsor **takeback** programs that recover a product or its packaging at the end of its life cycle.

2.5.1 Eliminating the Unnecessary Use of New Materials

The first step in greening building materials is to eliminate the unnecessary use of new materials. This means finding ways to get the job done with fewer materials. It also means substituting materials or parts of materials that aren't new whenever possible. Reusing materials is a green choice. Every product that is salvaged and reused saves energy and raw materials. **Recycled content** prevents the need to harvest **virgin materials**. This helps conserve valuable resources.

Perhaps the greenest choice of all is to reuse or adapt an existing building instead of building a new one. Not only do you eliminate the need for many new materials, you also reduce site development and infrastructure costs. Be aware that existing buildings present a challenge, as certain conditions are not always as expected. Sometimes contamination from lead-based paint, asbestos, or other substances must be remediated. Some systems may need to be replaced or upgraded. Avoid reusing windows, older HVAC equipment and appliances, and older plumbing fixtures. They are likely to be very inefficient and may no longer meet current standards.

Pay attention to packaging when procuring raw materials. In some cases, you can specify that products be delivered with reusable packaging that can be returned to the manufacturer. In other cases, packaging is **recyclable** or even biodegradable. If product packaging is **bio-based** (that is, made from substances derived from living matter), look for opportunities to compost or chip it on site for use as a soil amendment. This also reduces the need to purchase new materials for this purpose.

Salvaged materials can be a very green contribution to a project. If an existing building is being demolished, look for ways to reuse its materials in the new building. Often, concrete and masonry rubble can be reused for fill, subbase for pavements, or drainage. Timber in good condition can often be reused for structural purposes, or remilled for flooring or siding. Masonry units may be reused as well. Materials that cannot be reused on the new project may be worth salvaging for other projects. These may include doors, hardware, and various fixtures.

New materials with recycled content also reduce the use of virgin materials. Common materials with recycled content include steel and concrete. Recycled materials include a wide variety of finishes, structural materials, and even landscaping materials (*Figure 30*). With materials like steel and concrete, you may not even be able to tell the difference from virgin materials. Other products may include recycled content as composites. **Recycled plastic lumber (RPL)** is one such product. RPL combines **post-consumer** or post-industrial plastic waste with wood fibers or other materials. The result is an extremely durable wood substitute.

Recycling reduces the amount of waste sent to landfills and creates a source of raw materials for new products. However, not all recycling is the same. True recycling means using the waste to create the same kind of product, such as recycled

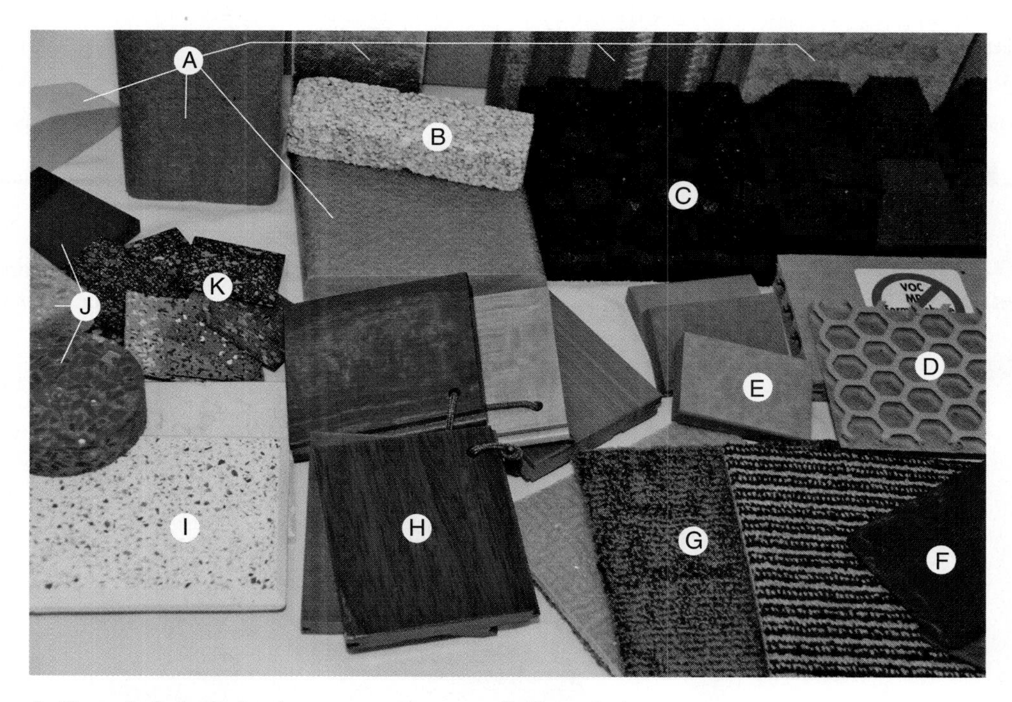

A. Recycled plastic lumber
B. Insulating concrete form material with recycled polystyrene pellets
C. Recycled rubber turf stabilizer
D. Recycled paper fiber board
E. Recycled paper fiber countertop
F. Recycled rubber shingles
G. Recycled content carpet (face fiber and backing)
H. Salvaged lumber
I. Recycled paper and glass cement tiles
J. Recycled glass countertops
K. Recycled rubber flooring

101F30.EPS

Figure 30 ◆ Waste-based and recycled content materials.

aluminum cans. Many other products can only be **downcycled**. They are turned into something else that can never be turned back into the original product. One example is plastics recycling. Recycling plastic into clear new containers is difficult due to mixing colors, degrading chemical bonds in the plastic, and contaminating the waste stream. Most recycled plastics are turned into composite products such as recycled plastic lumber, lower-grade plastic bags, or even carpet backing. Recycling plastics does reduce the amount of new materials required, but at a loss of both purity and quality. The best way to be green is to reduce the use of these products in the first place.

Recycled content can come from a variety of sources. Some recycled content is produced as a byproduct of manufacturing processes. This material is called **post-industrial/pre-consumer** recycled content. It can be recovered and reused without ever leaving the factory. Other recycled content comes from products that have been produced and used by consumers, then recovered for recycling. This material is called post-consumer recycled content. Post-consumer content is better than post-industrial recycled content because it truly closes the material loop. Post-industrial recycled content represents waste or inefficiency in manufacturing. Rather than recycling this material, it would be better to improve the manufacturing process.

Pollution prevention is the careful design of products and processes to eliminate the use or waste of materials. One example is to eliminate finishes and expose structural materials instead. For example, concrete floors can be stained, polished, or textured to create beautiful surfaces that do not require additional finishes, such as wood or tile. Exposed ceilings (*Figure 31*) eliminate the need for extensive dropped ceiling systems and provide visual interest while saving costs.

In many projects, natural systems can be used to perform the functions traditionally provided by engineered systems at lower cost. One example is living machines made of plants, snails, bacteria, and other organisms. Living machines, natural drainage swales, and constructed wetlands can be used to collect and treat wastewater without requiring any input of chemicals. The result is healthy plants and purified water that can be safely used for other purposes. Some of these practices cost more than conventional systems. However, you can often save enough money to pay for the investment through savings in other systems and/or life cycle cost savings.

Many building strategies also take advantage of the lessons offered by the green environment.

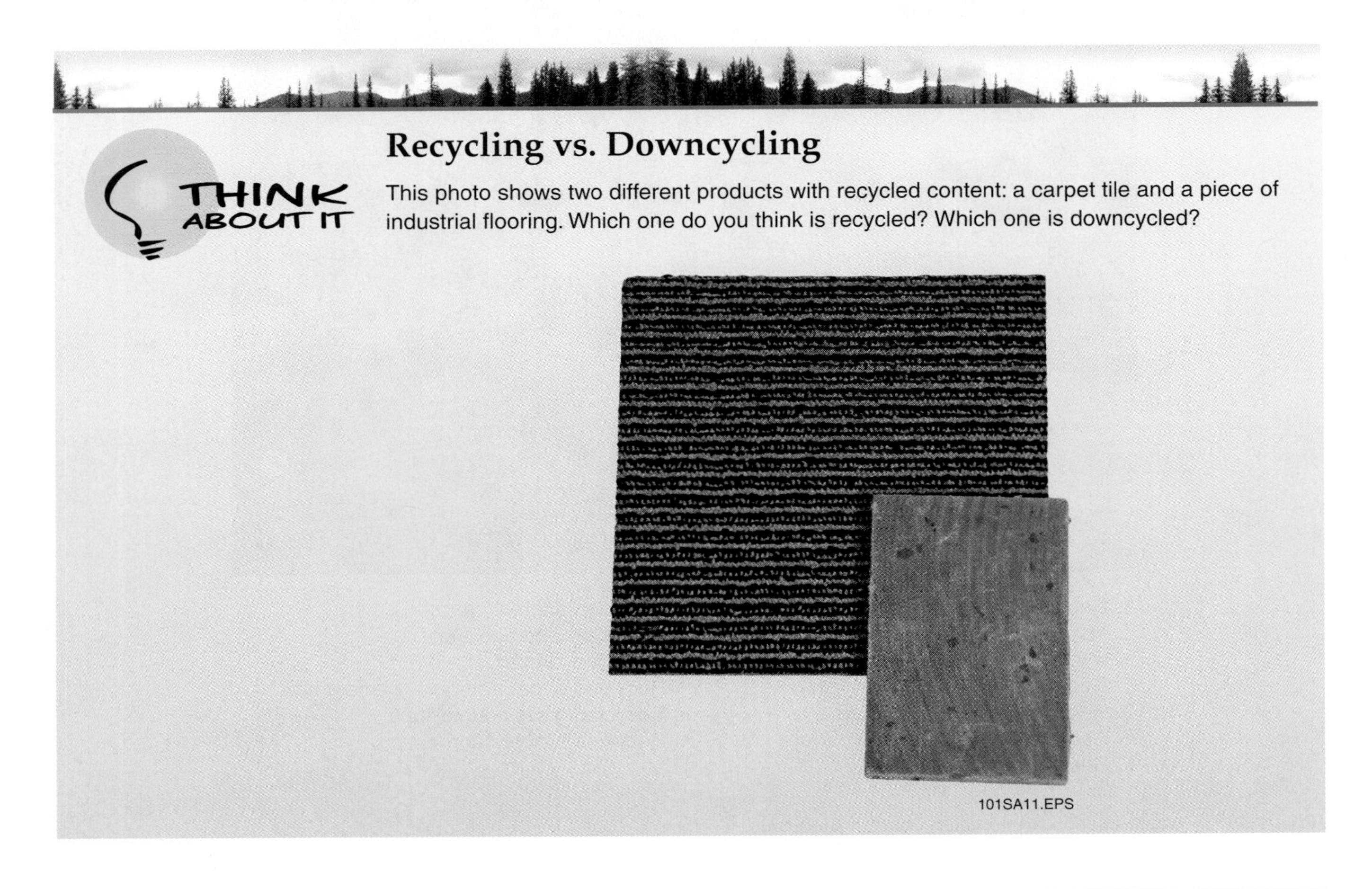

Recycling vs. Downcycling

This photo shows two different products with recycled content: a carpet tile and a piece of industrial flooring. Which one do you think is recycled? Which one is downcycled?

101SA11.EPS

Figure 31 ◆ Exposed ceilings eliminate unneeded materials.

This approach is known as **biomimicry**—imitating nature to create new solutions modeled after natural systems. One example is African termite mounds that maintain constant internal temperatures even with large fluctuations in outdoor air temperature. These mounds use an intricate network of tunnels to control the flow of air for optimum quality, temperature, and moisture levels. These self-sufficient mounds also include a type of farming. The termites supply a type of fungus with chewed wood fiber, which the fungus then breaks down into a usable food source. Nature produces solutions with optimum efficiency because plants and animals that cannot compete simply do not survive. As the world heads into an age of increasingly scarce resources and more people who need them, biomimicry offers valuable lessons for buildings and technologies alike.

2.5.2 Using Materials More Efficiently

The second strategy for greening materials is to use materials more efficiently, which means getting more benefit from the materials you do use and/or using fewer materials to achieve the same result. One way to do this is to use multi-function materials. Multi-function materials do more than one thing as part of a building. They can speed up construction, save building materials, and reduce waste. Every system and technology used in a project comes with overhead. This overhead includes extra packaging, transportation, and other costs that do not add value to the product itself. When one product serves as multiple systems, the overhead for that product is a fraction of the overhead associated with the multiple systems it replaces.

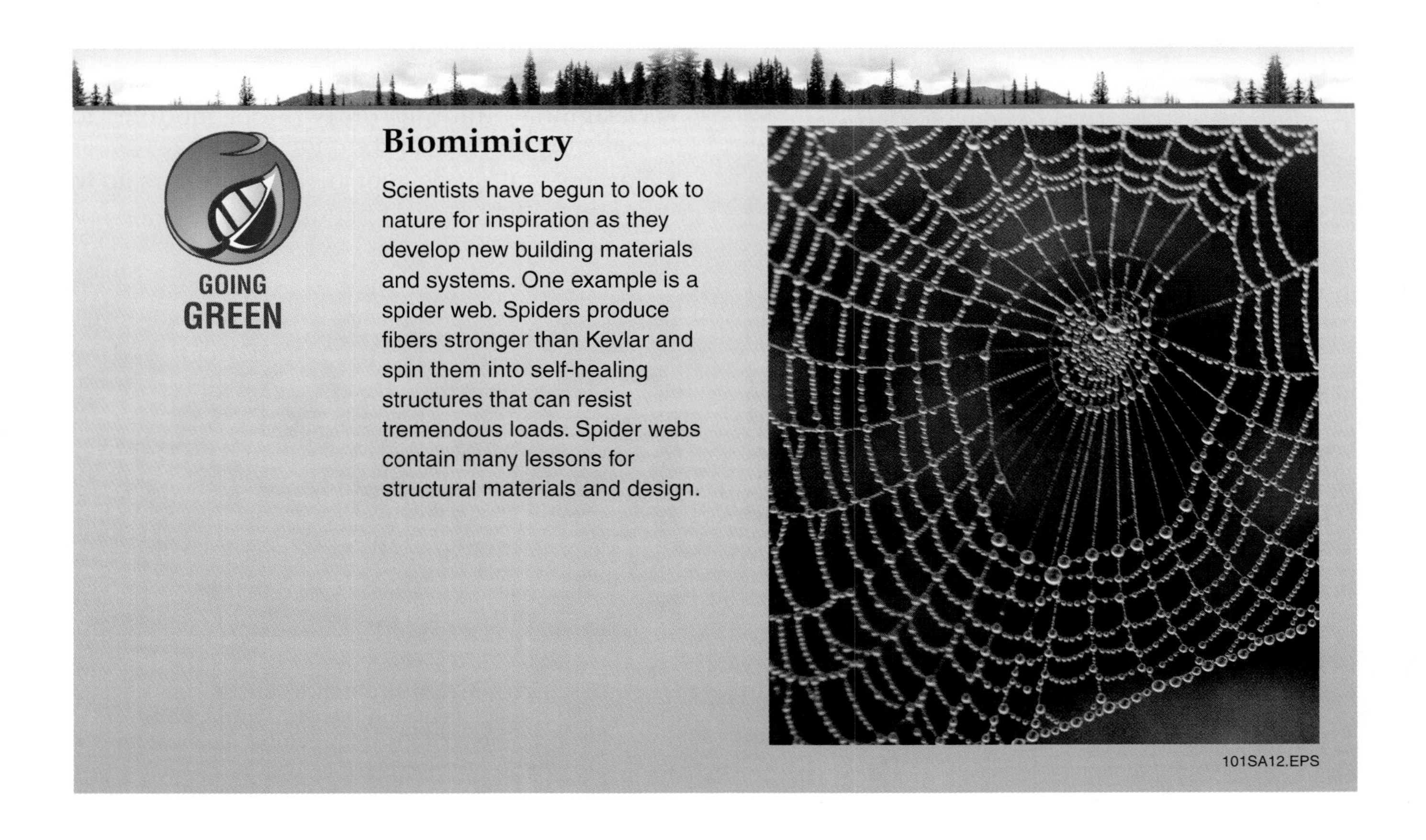

Biomimicry

Scientists have begun to look to nature for inspiration as they develop new building materials and systems. One example is a spider web. Spiders produce fibers stronger than Kevlar and spin them into self-healing structures that can resist tremendous loads. Spider webs contain many lessons for structural materials and design.

Many multi-function materials serve as the primary structure of the building. These include **insulating concrete forms (ICFs)**, **structural insulated panels (SIPs)**, and **aerated autoclaved concrete (AAC)**. Each of these structural systems (*Figure 32*) also serves as the building enclosure, insulation system, and mounting surface for interior and exterior finishes. By using these products, construction of one system achieves as much as three or four traditional building systems. This saves time, labor, packaging, transportation, and money.

An energy-related example is **building-integrated photovoltaics (BIPVs)**. These solar panels are built into components that play other roles in a building. Some BIPVs are used to generate electricity when the sun shines on a window. Other BIPVs are manufactured as solar shingles—they generate power while acting as the roof itself. Still others are used as shading devices for windows or parking lots.

A. Embedded steel stud SIP
B. Polystyrene insulating concrete form
C. Polystyrene stress skin SIP
D. Aerated glass stress skin SIP
E. Aerated autoclaved concrete
F. Fiber-reinforced aerated concrete
G. Straw core stress skin SIP

101F32.EPS

Figure 32 ◆ Multi-function materials.

Green roofs are another example (*Figure 33*). These vegetated roofs have energy benefits as well as benefits for roof life.

Green roofs are typically installed on top of a roof membrane. They can be of varying complexity, and help to ballast the roof itself. They also protect the membrane from solar radiation, which would otherwise cause the material to break down. They stabilize temperatures on the roof, reducing stress on the roof membrane and keeping the building cooler. They help to absorb rainfall, which reduces stormwater runoff. Finally, they are a microhabitat for birds, insects, and plants, and they help to clean the air. This single system, although it may cost more than a traditional roof, provides significant benefits that can make it a good investment. All benefits and costs must be considered when deciding what types of systems to use.

Optimal value engineered (OVE) framing for light wood frame construction is another way to use materials more efficiently. OVE framing is a good design solution because it allows more insulation to be incorporated in the walls of the building (*Figure 34*). This reduces the amount of heat that can escape through the walls. It also increases the energy performance of the building.

OVE framing also saves considerable materials during construction by using wood more efficiently. This means less lumber, fewer cuts, and less waste. According to the U.S. Department of Energy, key principles of OVE framing include the following:

- Designing buildings on two-foot modules to make the best use of common sheet sizes
- Spacing wall studs, roof joists, and rafters up to 24" on center

101F33.EPS

Figure 33 ◆ Green roofs provide enclosure, stormwater management, and thermal control while cleaning the air and providing a habitat for birds and insects.

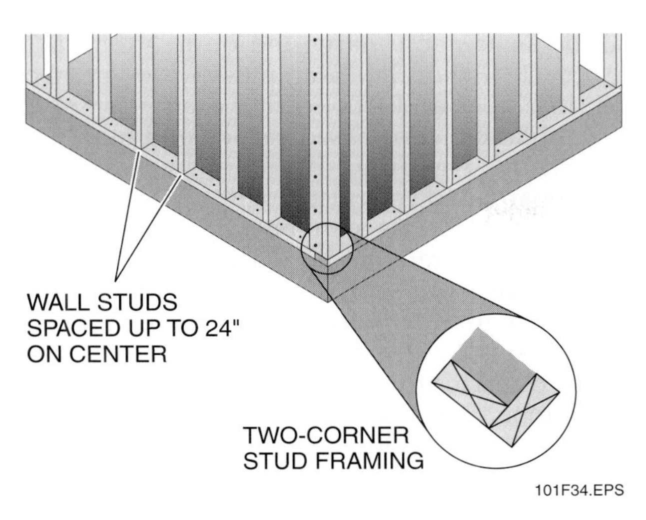

Figure 34 ◆ Framing details for OVE wood frame construction.

- Using in-line framing where floor, wall, and roof framing members are vertically in line with one another and loads are transferred directly downward
- Using two-stud corner framing and inexpensive drywall clips or scrap lumber for drywall backing
- Eliminating headers in walls that are not load-bearing
- Using single lumber headers and top plates when appropriate

These techniques reduce the amount of wood used and create a better thermal envelope. They also make construction easier for crews that must install plumbing, electrical, and HVAC services in the walls. Fewer structural members mean less drilling during installation and less effort to install. All of these attributes make OVE framing a good strategy to green your building and save both first costs and life cycle costs.

Another approach is to use **smart materials**. Smart materials work by changing in response to environmental conditions. For example, smart windows may automatically increase their level of tint in response to higher levels of sunlight. Smart window blinds may automatically adjust to follow the sun's path. Some materials are under development that can change colors or other properties as well.

A special class of smart materials is called **nano-materials**. Nano-materials use extremely small particles that add functionality to building materials. For example, nano-coatings can be used to make paint repel dirt or concrete absorb air pollution. Nano-coatings are also being used to add mold resistance or biocidal (germ killing) properties to finishes. A new generation of photovoltaic panels are being developed using carbon nano-tubes. These nano-tubes improve efficiency by increasing the surface area exposed to the sun.

New generations of lightweight modularized construction systems are now available that can also do more with less. Prefabricated, factory-assembled building components reduce waste. They use materials more efficiently since they are manufactured in a controlled environment. Regular production allows manufacturers to optimize the use of materials and produce the most efficient components possible.

Modular construction components include carpet tile (*Figure 35*), raised floor systems, and demountable furniture systems. These systems allow rapid reconfiguration to meet changing user needs. They also permit replacement on a unit-by-unit basis in the case of damage. This means avoiding the need to replace entire rooms of carpet for one stained or worn area. Even regular materials can be used for modular construction. Use regular dimensions that match the standard sizes of building materials. For example, don't design decking that is 9 feet long. Decking comes in increments of two feet. Expand your deck by another foot and avoid having to waste material.

Choose systems and materials of appropriate durability. If you know that the facility you're building is going to be remodeled every 5 to 7 years, don't use granite countertops unless you plan to recover and reuse them somewhere else.

Figure 35 ◆ Carpet tile is a modular material that prevents waste.

This idea is known as design for disassembly. Put your systems together in a way that you can take them apart at the end of their service life for possible reuse.

During the construction process, use lean construction methods to maximize efficiency. Lean construction eliminates unnecessary steps and materials, and eliminates waste by changing the way things are built. For example, **just-in-time delivery** of materials to the site means that products are installed right off the truck. They do not need to be stored or staged. This eliminates the opportunity for materials to become damaged in storage. It also minimizes the opportunity for loss or damage as products are moved around the site. Careful material tracking helps to avoid waste. A new method of material tracking involves the use of **radio-frequency identification (RFID)** tags. These tags make it easy to locate materials in crowded staging areas. If materials must be stored before installation, protect them from moisture and possible damage. Centralized cutting operations also help to increase material efficiency. Keep cutoffs in the the same area where all cutting is done. This makes workers more likely to use them instead of pulling a new piece of material from the supply stack.

2.5.3 Finding Better Sources of Materials

Another way to green your use of materials is to find better sources. For example, **rapidly renewable** materials have become common. Rapidly renewable materials are a special class of bio-based materials that grow quickly. According to the U.S. Green Building Council (USGBC)®, rapidly renewable materials are any material that can be sustainably harvested on less than a 10-year cycle (*Figure 36*). Examples include bamboo, cellulose fiber, and wool. Other examples are cotton insulation, corn-based carpet, and blown soy insulation. Agrifiber, linoleum, wheatboard, strawboard, and cork also qualify.

Other materials can be considered green because they are abundant. This includes buildings made from soil, such as **rammed earth**, adobe, or **cob construction**. Cob construction is an ancient building method using hand-formed lumps of earth mixed with sand and straw. Like cob, adobe typically includes plant fibers to help stabilize the soil. Other earth-based construction methods strengthen the soil by using small quantities of cement or lime. An innovative material in this category is **papercrete**. This material is a fiber-cement that uses waste paper for fiber.

Strawbale construction uses both abundant and bio-based materials. The bales of straw used for this purpose are typically three to five times more densely packed than agricultural bales. Strawbale construction can be either structural or used as fill with a wood frame or other structure. Bales are stabilized using rebar or bamboo spikes. Bond beams are used to ensure a load-bearing surface along the tops of walls. Strawbale is becoming more widely used in certain parts of the United States. Arizona even has a special section of the building code for this construction method.

Proper detailing to prevent moisture problems is critical for success. Deep overhangs and moisture barriers between bales and foundations are common. Strawbale construction has a high insulating value. It also has a two-hour or greater fire rating depending on the exterior and interior finish.

Recycled content materials are also a better source than virgin materials for many applications. Post-consumer recycled content is especially desirable. However, many recycled content materials are composites. This means that they are combinations of multiple different types of raw materials. The prospects for recycling composite materials are limited. However, some composite materials outperform conventional materials. They may require lower maintenance and fewer raw materials. People often confuse the term recyclable with recycled, but there's a big difference between the two. Recyclable means that the product itself can be recycled after it is used. Recycled means that a product contains material that has been previously used, either in industry during manufacturing (post-industrial) or by consumers (post-consumer). Post-industrial recycled content is less desirable than post-consumer recycled content. This is because post-industrial recycled

What's Wrong with This Picture?

101SA13.EPS

A. Soy-based foam insulation
B. Wool carpet
C. Corn-based carpet fiber
D. Coir/straw soil stabilization mats
E. Corn-based carpet backing
F. Coconut wood flooring
G. Wheat straw fiberboard
H. Sorghum board
I. Bamboo flooring/plywood
J. Linoleum
K. Hemp fiber fabric
L. Cork flooring
M. Cotton fiber batt insulation

101F36.EPS

Figure 36 ◆ Bio-based and rapidly renewable materials.

content often results from wasteful manufacturing processes that could be improved. Scientific Certification Systems is an organization that provides certification of the recycled content of materials. Look for the SCS logo on products to be sure that their recycled content has been verified (*Figure 37*).

Look for sustainably harvested materials. The Forest Stewardship Council (FSC) is an organization that evaluates wood products to determine if they are sustainably harvested. Look for **FSC Certified** wood products (*Figure 38*). More information on the FSC certification standards is available online at www.fsc.org.

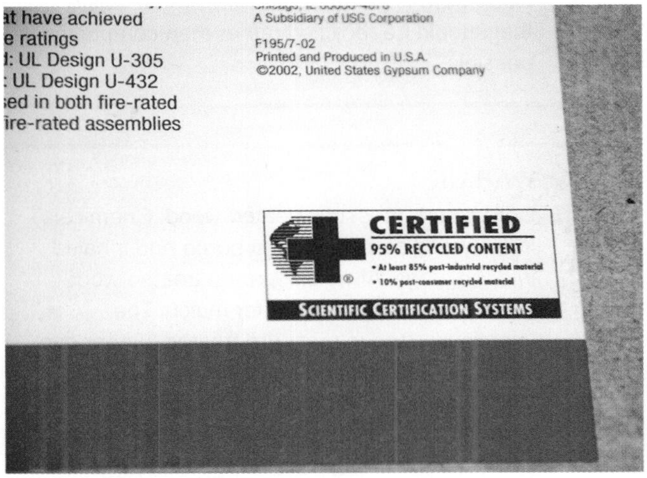

101F37.EPS

Figure 37 ◆ SCS Certified content label.

101F38.EPS

Figure 38 ◆ FSC Certified wood product.

Seek out local sources of materials wherever possible for your projects. Using locally produced materials reduces the energy required to transport them. This is one of the biggest impacts of building material. Often, locally harvested materials can be obtained at lesser cost than imported materials. Using local materials can save time for procurement and helps to support the local economy.

2.5.4 Finding Better Sinks for Waste Streams

The final step you can take to green your materials use is to find better uses for building waste. Depending on the type of project you are building, multiple options may be available. Consult salvage companies for projects that involve the removal or rehabilitation of existing buildings. This will ensure that all reusable materials are removed before demolition begins. Older buildings may be good candidates for deconstruction. Deconstruction is the systematic disassembly of a building in the reverse order from how it was built. This process salvages timber and other materials for reuse.

Both salvaged materials from existing buildings and leftover materials from new projects can be donated to local nonprofit organizations involved with construction. Habitat for Humanity (www.habitat.org) is a nonprofit organization that operates ReStores in various cities around the country (*Figure 39*). ReStores are retail outlets that sell donated building materials. Companies that donate materials to a ReStore can take a tax deduction for their charitable contribution. At the same time, the ReStore obtains a source of materials that it can sell to raise money to build homes for needy families. If your project is likely to have usable materials left over, contact your local ReStore to ask about on-site pickups and donation opportunities. You can also shop at the ReStore yourself if you need small quantities of a certain item or would like to find salvaged or used materials for your next project.

101F39.EPS

Figure 39 ◆ Habitat for Humanity ReStores accept donations of construction materials.

For waste that is not suitable for donation, consider recycling as an option for disposal. Metal recycling is available throughout the United States. Organizations that recycle cardboard, drywall, ceiling tiles, carpet, concrete, masonry, and asphalt shingles are becoming more common, especially in major metropolitan areas. Consider contracting with a waste recovery company to simplify recycling. Depending on the waste streams generated, these companies deliver dumpsters at different times during the project. They may also provide worker training or housekeeping services to ensure that materials are sorted correctly. Some areas also have facilities for off-site sorting of co-mingled construction and demolition waste. These facilities take your waste at a reduced rate and sort it to recover recyclable materials. Depending on market conditions and the composition of your waste stream, they may even pay you for your waste. In addition to typical building materials, batteries can also be recycled. **Lithium ion (Li-ion)** and **nickel cadmium (NiCad)** batteries are both rechargeable and recyclable.

Bio-based waste is a good candidate for on-site chipping, mulching, or composting. Pallets, wood waste, cardboard, and even drywall can be chipped or mulched on site by specialty contractors. The result is a product that can be used as a soil amendment or for landscaping purposes.

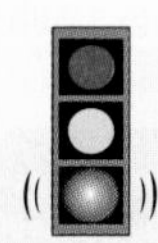

NOTE

Cardboard is generally a higher value waste that should be recycled rather than composted if possible.

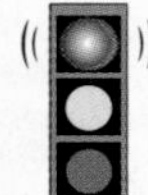

WARNING!

Do not burn pressure-treated wood. Chemicals in the wood can become airborne and inhaled into the lungs. Always separate treated wood from wood being chipped for mulch. The chemicals used in pressure-treated lumber can leach into the soil and contaminate produce. Collect scraps and place in a dumpster reserved for pressure-treated material. Wash your hands thoroughly after handling pressure-treated material.

Some contractors also grind concrete and masonry rubble for use on site as a fill or subbase.

Certain kinds of waste, if produced in large quantities, may be good candidates for storage in a **monofill**. Monofills are landfills that take only one type of waste, typically in bales. Monofills are designed to store waste so that it can be easily recovered later. Most waste destined for monofills is high-value waste. Cost-effective recycling methods are now being developed for wastes in monofills. A common type of monofill is carpet waste. Recent advances in separation technology have made carpet monofills into mines for raw materials.

As the world population continues to grow, **overpopulation** is a concern. It can lead to overcrowding and the depletion of natural resources. Smart use of materials for construction can help ensure an ongoing supply of materials to meet human needs both now and in the future.

2.6.0 Indoor Environment Best Practices

Most people spend over 90 percent of their time indoors in climate-controlled buildings. These buildings are sealed against leaks to maximize energy efficiency. At the same time, the materials used to build these structures have become largely composite and/or synthetic. From carpets to engineered wood products, much of the indoor environment now consists of products that emit chemicals over their life spans.

Together, these factors have led to a sharp increase in building-related health symptoms known as Sick Building Syndrome. Symptoms include headaches, fatigue, and other problems that increase with continued exposure. With increased concerns for energy conservation, building operators often reduce ventilation rates in unoccupied spaces or during nights and weekends. Reduced ventilation rates, along with tighter building construction and chemical emissions from building products, greatly reduce indoor air quality. In addition, many products such as drywall, carpet, and ceiling tiles serve as excellent food sources for mold. Inadequate ventilation, moisture, and food sources have led to explosions of mold growth in some construction projects. This has had disastrous results for building occupants.

People also contribute to the problem by virtue of their activities. The basic act of breathing increases CO_2 levels in a room. Higher concentrations of CO_2 can lead to drowsiness, reduced productivity, or even headaches and discomfort. Activities like smoking, cooking, or using printers and photocopiers contribute pollutants to the indoor air. Shedding skin cells and the gases produced from digestion also contribute to poor indoor air quality.

The indoor environment is critical for proper building operation. Keeping occupants happy, healthy, and productive is essential for good business and a happy home life. This section explores ways to green buildings by improving indoor environmental quality. The first tactic is to prevent potential problems at their source. Next, for pollution sources that cannot be avoided, take measures to isolate the source of pollution to prevent it from spreading. Third, use the laws of physics and psychology to create pleasant and well-functioning spaces for building occupants. Finally, give users control over their environment to allow them to make adjustments for comfort and productivity.

2.6.1 Preventing Problems at the Source

Often, decisions made during the design of a building can have a huge impact on indoor environmental quality. For example, locating buildings upwind of major pollution sources such as power plants can eliminate the need to try to fix air quality later. The same applies to locating buildings away from noise sources such as highways. Careful placement of air intakes away from pollution sources such as loading docks can also prevent problems. Be sure all vents and exhaust areas are located downwind from air intakes. Consider the features of neighboring buildings when deciding where to locate air intakes and exhausts.

The materials and activities used to construct a facility have the potential to create problems if not carefully selected. Many modern finishes such as paints, sealants, carpets, and composites contain solvents and adhesives that release VOCs as they age. The **offgassing** of VOCs is an example of **fugitive emissions**, which means they are not from a stack or vent. That new car smell that many people like is actually caused by the materials releasing VOCs, such as **urea formaldehyde**. If you can smell an odor, the product that caused it is already on its way to your lungs. Select products with a low or zero VOC content. Most major paint producers now have lines of low VOC paints that perform as well and cost about the same as traditional paints. These paints are **water-based** rather than **solvent-based**. The only limitation of these products is that you can't mix dark colors. The pigments required to achieve extremely dark colors contain VOCs. Look for the Green Seal logo on paints to ensure that they meet the criteria for low VOCs (*Figure 40*). **Green Seal Certified** paints and sealants are listed online at www.greenseal.org.

Carpets can also be certified for indoor air quality using standards developed by the Carpet and Rug Institute (CRI). The CRI's Green Label Plus logo is an indication that a carpet has been tested and emits minimal levels of VOCs (*Figure 41*). A list of current Green Label Plus certified products is available at the CRI's website at www.carpet-rug.org. Carpets and rugs are recertified quarterly to ensure that they continue to meet these rigorous standards.

Figure 40 ◆ Look for the Green Seal Certified paint logo.

Figure 41 ◆ The Carpet and Rug Institute's Green Label Plus logo for certified carpet.

Proper design, drainage, and landscaping can help prevent water intrusion into a building. Wherever water or moisture exists at warm temperatures with an available food source, mold or mildew is likely to grow. This can be prevented through good design, ventilation, and maintenance. Make sure the areas surrounding the building are graded properly to drain water away from the building. This helps prevent moisture problems in basements and crawl spaces. Crawl spaces should also have a **waterproof** barrier between the soil and the conditioned space.

Careful selection of landscape plants can also help to prevent problems. Many ornamental trees and shrubs produce a significant amount of pollen. So do species such as olive, acacia, oaks, maples, and pines. Pollen produced by these trees can aggravate allergies. It can also contaminate air intakes and rainwater harvesting systems. Ask about the pollen production of landscape plants before you make your final selection.

During construction, pay attention to the sequencing of construction activities. Schedule dust-producing and dirty activities before carpets or other **absorptive finishes** are installed. Other absorptive finishes include wall coverings, fabrics, ceiling tiles, and many types of insulation. Ensure that materials are stored on-site in a way that keeps them safe from moisture and potential mold growth. If possible, use just-in-time delivery to minimize the amount of time materials sit around the site. Never install material that has become wet without testing to be sure it will not become contaminated with mold. Wait until after the building shell is complete with windows, doors, and a roof before installing any finishes or materials that can retain moisture. Be sure to allow adequate time for curing of concrete and finishes and provide adequate ventilation. Significant moisture is released during curing. If it has nowhere to go, it will remain in the building and cause problems.

Select finishes that are nonabsorptive such as tile floors and painted walls (*Figure 42*). These finishes are easier to clean and **water-resistant**. They also do not trap particulates and odors that can cause occupant discomfort. Choose finishes that contain biocidal (germ-killing) or mold-resistant coatings. Natural linoleum has biocidal properties, which makes it a good choice for flooring.

After the building is finished, green housekeeping practices can help maintain a healthy indoor environment. Follow manufacturer's instructions for maintaining finishes. For example, vacuum carpets often to prevent the buildup of irritants such as dust mites. Read labels on cleaning chemicals. Choose **nontoxic**, water-soluble cleaners wherever possible.

101F42.EPS

Figure 42 ◆ Nonabsorptive finishes such as ceramic tiles are easier to clean and do not absorb pollutants.

Monitor building humidity and maintain relative humidity levels between 30 percent and 50 percent in all occupied spaces of the building. During summer, cooling coils may require maintenance to ensure that they are properly dehumidifying incoming air. Inspect and properly maintain all HVAC equipment. This enables you to identify and fix problems before they create unhealthy conditions.

2.6.2 Providing Segregation and Ventilation

The second step in achieving good indoor environmental quality is to segregate activities and materials that create pollution from other parts of the building. Segregation is especially important during construction activities to prevent contamination of building systems. Use separate ventilation systems for high-risk areas. Entrance control measures such as **walk-off mats** help block contaminants that can enter the building through entryways.

Any type of construction activity has the potential to cause future indoor air quality problems if not properly managed. Many construction activities, such as cutting operations and sanding, produce dust or airborne particulates. Other activities involve installing materials that release VOCs or moisture while they cure. These materials include paint, sealants, adhesives, concrete, and new lumber. Even with dust collectors installed on equipment, building systems must be protected from absorbing pollutants during construction. These contaminants reduce indoor air quality and may result in damage to sensitive building systems. Proper protective measures reduce cleanup time, which speeds construction and lowers the cost (*Figure 43*).

At a minimum, construction isolation measures should include protection of all HVAC intakes, grills, and registers. Use plastic sheeting and duct tape or ties to seal off all possible points of entry to building ductwork or plenums. Be sure to deactivate the HVAC system before you do this. Also, isolate entryways to other parts of the building and stairwells or corridors outside the work areas.

Install absorptive materials after dust-producing activities are completed. Absorptive materials include carpet, wall coverings, fabrics, ceiling tiles, and many types of insulation. If this is not possible, use plastic or paper sheeting to protect absorptive surfaces.

Maintain proper ventilation during construction to ensure worker health and safety by exhausting pollutants from the workspace. If possible, maintain continuous negative pressure exhausted to outside air. Exhausting workspace air prevents dust and contaminants from migrating to other parts of the building. Ventilation should be provided by a temporary ventilation system to avoid contaminating building ductwork with pollutants.

Regular housekeeping during construction keeps contaminants under control and prevents them from becoming airborne. Specialty contractors can be retained for this purpose. Alternatively, work practices can be put in place to require workers to keep their areas clean and free of debris. All construction isolation measures should be regularly inspected for leaks and tears and replaced as needed.

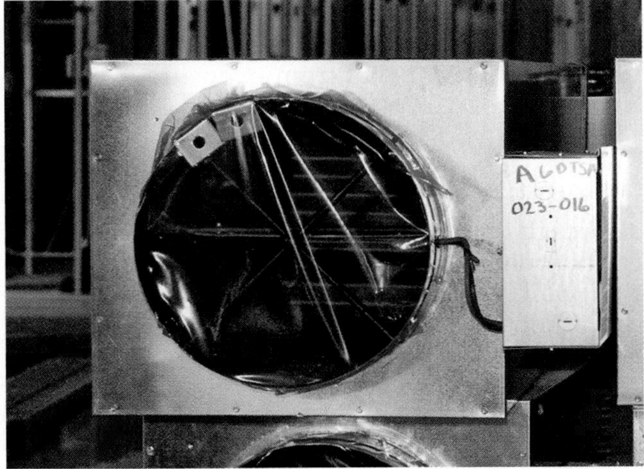

101F43.EPS

Figure 43 ◆ Construction isolation measures ensure that contaminants from construction tasks do not contaminate building systems.

At the end of construction, remove all masking and sheeting used to isolate ventilation systems. Temporary high-efficiency filters may be used in the HVAC system during some construction activities and before building occupancy. The efficiency of a filter is measured by the **minimum efficiency reporting value (MERV)** scale. High-efficiency filters trap remaining contaminants and prevent them from entering the system. Replace all filters prior to occupancy. The EPA recommends a two-week **flushout** period for all new buildings prior to occupancy. During flushout, ventilation systems are run continuously with full outdoor air. This accelerates the offgas removal from building materials. However, it can also introduce large amounts of humidity into the building depending on the climate and season. This humidity adds to the already high humidity rates from the curing of materials inside the building itself. If a building flushout is undertaken, ensure that humidity controls are in place to handle the extra humidity load that may occur.

Research has shown that the primary sources of household pollutants are ordinary products such as paint, cleaning compounds, personal care products, and building materials. Everyday tasks such as bathing, laundering, cooking, and heating can all contribute to poor indoor air quality.

Areas of concern in commercial buildings include kitchens, office equipment rooms, housekeeping areas, and chemical mixing and storage areas. Provide separate ventilation for these areas to minimize the risk of pollutants spreading to other parts of the building. In addition, these areas may be isolated from other spaces in the building through full deck-to-deck partitions and sealed entryways, depending on the risk level of contamination. Indoor smoking areas are special cases that require careful design and construction to protect nonsmokers from exposure to second-hand smoke.

Entrance control is important to keep dirt and pollutants outside the building from getting inside. Air lock entryways and vestibules are one approach that can also save energy by keeping conditioned air inside the building and unconditioned air out. Entry areas can benefit from the use of walkoff mats and dirt collectors, particularly in areas where inclement weather may lead to snow and ice being tracked into the building. Walkoff mats are also available for use outside construction areas to help prevent contamination of other areas of the building.

2.6.3 Taking Advantage of Natural Forces

A variety of natural forces can affect the indoor environment. The laws of physical science can be used to provide natural lighting, ventilation, heating, and cooling in a space. This not only saves energy, but also can provide better indoor environmental quality for occupants.

Paint colors are one of the easiest ways to influence the experience of a space (*Figure 44*). Often, they require no additional expense. Color can influence both how you perceive spaces and how you feel about them. Dark colors can make work-spaces feel cramped and depressing, and tend to make rooms look smaller and more confined. Lighter colors create a sense of openness, may improve mood, and make rooms look larger and more spacious. Dark colors make high ceilings look closer, while lighter colors make low ceilings look higher. Combinations of colors can be especially effective in creating focus areas and drawing people into a space.

Natural daylight can be used to enhance the indoor environment in many ways. It can also cause problems if not managed well. Potential problems include excessive heat gain, fading, and glare. Light shelves or louvers are one way to incorporate daylight into a space without overheating spaces along the building perimeter. Light shelves bounce sunlight onto the ceiling of a room and reflect it further back into the building space. In combination with reflective or light-colored ceilings, light shelves spread the benefits of natural light throughout a space.

101F44.EPS

Figure 44 ◆ Paint color choices can influence the user's experience of a building space.

DID YOU KNOW?

The Psychological Impacts of Color

Psychological research has shown that people react to different colors in predictable ways. Common colors and their psychological reactions are as follows:

- Yellow is highly visible and often used for safety markings. Yellow can make small rooms look larger and bring light to narrow entranceways and hallways.
- Orange is perceived to be cheerful and friendly. It is a good choice for rooms where people gather informally.
- Red stimulates the pituitary gland and raises heart rates and blood pressure. It also stimulates appetite. Red encourages action and aggressiveness. It is often used on buttons and knobs.
- Blue is a calming, soothing color. Blue works well in bedrooms.
- Green evokes feelings of relaxation and quietness. Green works well in study or work environments since it promotes concentration.
- Purples can reduce blood pressure and suppress appetite.
- Brown works well in environments where food is prepared or eaten. It also works well in general living environments. Brown is associated with comfort, reliability, and warmth.
- Gray encourages creativity. Gray interiors can be perceived as depressing unless accented with bright, clean colors.
- White is good around food and in precision work environments since it suggests sterility. Pure white can be perceived as harsh. Off-whites are better suited for many applications.
- Black is perceived to be dignified, sophisticated, and elegant. It tends to visually recede and enhances other colors used in combination with it.

Natural ventilation, when carefully designed, can also create a comfortable indoor environment with minimal energy. *Figure 45* shows an example of a natural ventilation system that would work well in an arid climate. Wind scoops are used to capture prevailing winds and pass them over a receptacle of moisture. As the air absorbs moisture, it becomes cooler and sinks down into the living space. As it mixes in the living space and becomes stale, its temperature gradually increases. The warmer stale air rises to ceiling level and is exhausted downwind with another air scoop. This type of system relies on absorption of moisture to cool the air. In climates with high humidity, other approaches are required.

Finally, geothermal heating and cooling takes advantage of the fact that the earth's temperature below the frost line stays relatively constant year round. Geothermal heating and cooling systems use pipes embedded in the ground or in nearby ponds or streams to shed heat in the summer and absorb heat in the winter. The fluid in these pipes is preheated or precooled by this contact with the earth and used to condition indoor environments. Geothermal heat pumps can also be used to heat hot water. They are very energy efficient, despite a relatively high initial cost.

2.6.4 Giving Users Control over Their Environment

Giving users control over their environment helps to improve indoor environmental quality for building occupants. Various methods can be used to achieve this. Operable windows, climate controls, and underfloor air distribution systems (UFADs) all provide occupants with the ability to control their individual spaces. The effects on building users can be significant. Studies have suggested that occupants who can control their spaces are happier, have greater job satisfaction, take fewer sick days, and are more productive.

Operable windows have gone in and out of vogue in architectural design over the past decades. Traditional buildings without mechanical heating and cooling had operable windows to allow users to control the indoor climate. Many contemporary buildings, however, have mechanical systems that don't work well with uncontrolled introduction of outside air. For this reason, many commercial and institutional buildings do not include operable windows. This creates a threat to **passive survivability** if the power supply to run mechanical equipment becomes unavailable. Passive survivability is the ability of a building to continue to offer basic function and

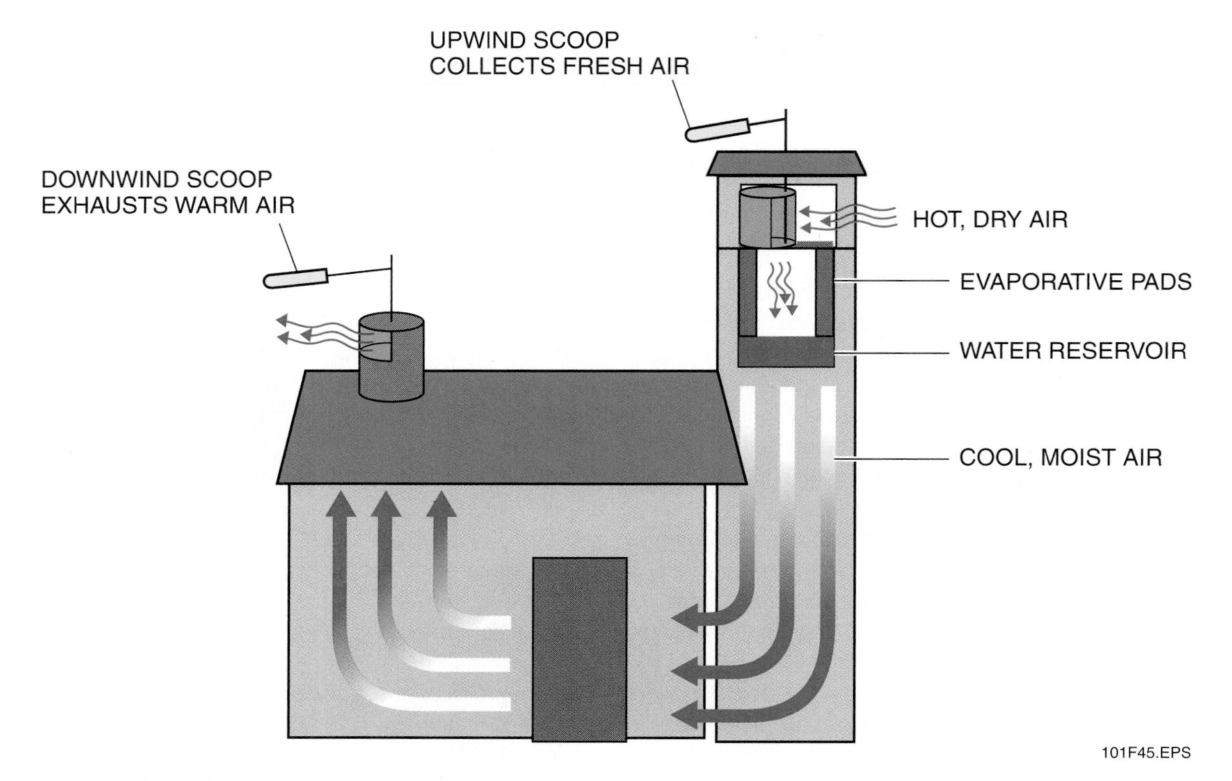

Figure 45 ◆ Natural ventilation system.

habitability when the infrastructure goes down. There is growing interest in designing buildings around this idea. Many current buildings operate poorly or fail when deprived of power. In many climates, operable windows can be used in lieu of mechanical ventilation during the swing seasons (fall and spring) when outdoor temperatures are mild. This can save considerable energy for operations. It also contributes to user satisfaction with individual workspaces.

Lighting, temperature, humidity, and ventilation controls are also becoming more sophisticated in modern buildings. Many commercial buildings rely on elaborate sensor and control systems to adjust these variables. However, each individual user of a building is different. Not all users are comfortable with the same environmental conditions. Individual controls at each workstation or office can help users adjust conditions to meet their individual needs (*Figure 46*). It can also increase their satisfaction with the space.

Underfloor air distribution systems are one way to increase the level of control users have over their individual workspaces. These systems provide the ability to control ventilation rates and perceived temperatures at each individual workspace. They also increase ventilation effectiveness and promote a uniform flow of conditioned air through the workspace (*Figure 47*).

Figure 46 ◆ Giving users control over their environment can increase productivity and satisfaction.

These systems can be combined with modular office furniture to allow rapid reconfiguration of space. They make it easy to adjust the configuration of space while maintaining proper ventilation and space conditioning in each work area. Modular control units can be moved around to provide airflow wherever users are located. This level of flexibility and control means that users are happier in their workspace. It also means they are less likely to need supplemental space heaters or fans to be comfortable.

DID YOU KNOW?

Passive Survivability

Hurricanes, tornados, floods, earthquakes, and a variety of human-created disasters or errors can cause the electricity or plumbing to go down at any time. Modern skyscrapers, such as the one shown here, typically have poor passive survivability. These buildings rely on electrical power for everything from ventilation to elevators to water pumps. Users may not even be able to open a window due to the building design. The principles of passive survivability align well with green building. Designing buildings to function in the absence of external supplies of energy, water, and materials makes them less vulnerable to external threats. It also means that they rely less on the environment for these resources. Passive solar design, natural ventilation, rainwater harvesting, durable materials, and graywater reuse are all tactics that both increase passive survivability and green a building at the same time.

101SA14.EPS

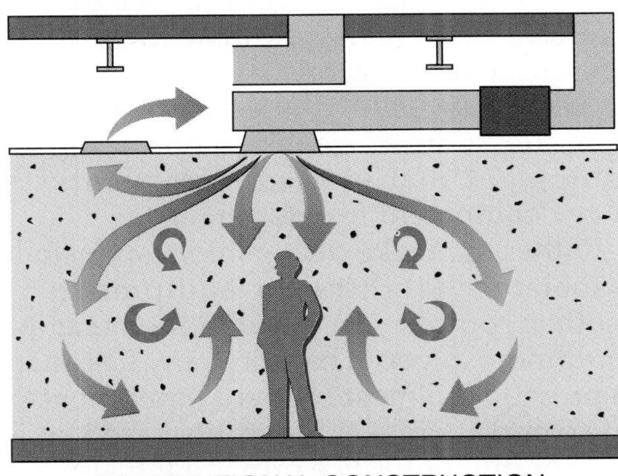

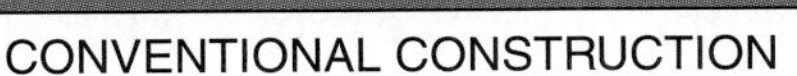

101F47.EPS

Figure 47 ◆ Underfloor air distribution systems help conditioned air reach all parts of a space.

2.7.0 Integrated Strategies

Many of the tactics and technologies described in the previous section offer multiple benefits for the green environment. The challenge for you is to find green solutions that meet the owner's needs for a facility, don't damage natural ecosystems or deplete resource bases, and don't cost more than a traditional project would cost. This section highlights strategies for green building decisions that achieve multiple benefits and affect multiple systems. Finding good integrated solutions can help you create better green buildings that cost less to build and have long-term benefits as well.

2.7.1 Solving the Right Problem

The first strategy for greening a project should be to make sure you're solving the right problem. Too often, building professionals leap into planning and designing a new facility without considering other options that might offer significant advantages. Possibilities include leasing, renovating an existing building, and telecommuting. In some cases, this could result in both green advantages and performance advantages at a much lower cost than building a new structure. In other cases, the most appropriate solution may still be to build new. However, considering other options early in the process can lead to a different mindset about the project that can help you be more creative on later decisions throughout the project.

The key to making sure you're solving the right problem is to focus not on solutions, but on the needs those solutions will meet. For example, how many times have you been asked at the checkout line whether you'd prefer paper or plastic bags? If you focus on these two choices as the only options available to you, you might miss other solutions. However, if you phrase the question as, "How can I transport my groceries from store to home?" you can consider other options entirely. What about reusable canvas bags, grocery delivery services, or even modular shopping carts that fold up to fit in your car? You might even decide to grow some of your own food to reduce your need for groceries. To use this principle, determine the functional necessity to be met. If you find yourself defining your problem in terms of one or more solutions, chances are you need to take a step back and reassess your options.

2.7.2 Exploiting Relationships Between Systems

The next way to use integrated strategies to green a project is to exploit the functional relationships between systems. Building systems are related to one another. The design of one system affects the design of another. For example, increasing the weight of a structure means that the foundation has to be increased as well. Sometimes these relationships can be exploited to pay for investments in one system through savings in another. This is known as **integrated design** (*Figure 48*).

For example, consider using high-performance windows as part of the building envelope design to provide for daylighting. Considered in isolation, this can raise total project cost. However, using these windows provides the benefits of better envelope performance and reduced heat load from light fixtures. Therefore, the extra costs of better windows can be offset by reducing the capacity of the building cooling system. In addition, a smaller HVAC system might mean smaller pumps, fans, and motors, reduced duct sizes, smaller plenums, and reduced floor-to-floor height, also reducing the cost of the facility. Reduced floor-to-floor height means less surface area of the building envelope, which means less material costs for the system. It also means that the overall weight of the building is reduced, meaning that foundations can be smaller and more efficient as well.

In the end, the overall increase in the total first cost of the project may be negligible if the benefits of improving one system are captured in the design of related systems. More importantly, life cycle cost savings can be even greater with these more efficiently designed systems. HVAC systems in particular will be much more efficient if they are correctly sized for the facility, allowing them to operate at maximum efficiency over the life cycle of the facility.

2.7.3 Using Services Rather than Products

A third strategy being used to green construction projects is dematerialization. Dematerialization refers to using services instead of products to meet user needs. Construction companies have successfully used this concept for years. For example, a small general contractor may rent equipment that is used infrequently, such as heavy equipment for grading.

Rather than take ownership responsibility (and associated liability) for equipment, you pay another company to provide the benefits of that equipment to you. That company has an incentive to provide the most efficient equipment possible. The company makes its profits based on a fee per hour or unit of service provided. It also has incentive to design products that can be easily repaired, upgraded, or disassembled. This is because the company retains responsibility for the ongoing maintenance and eventual disposition of the equipment.

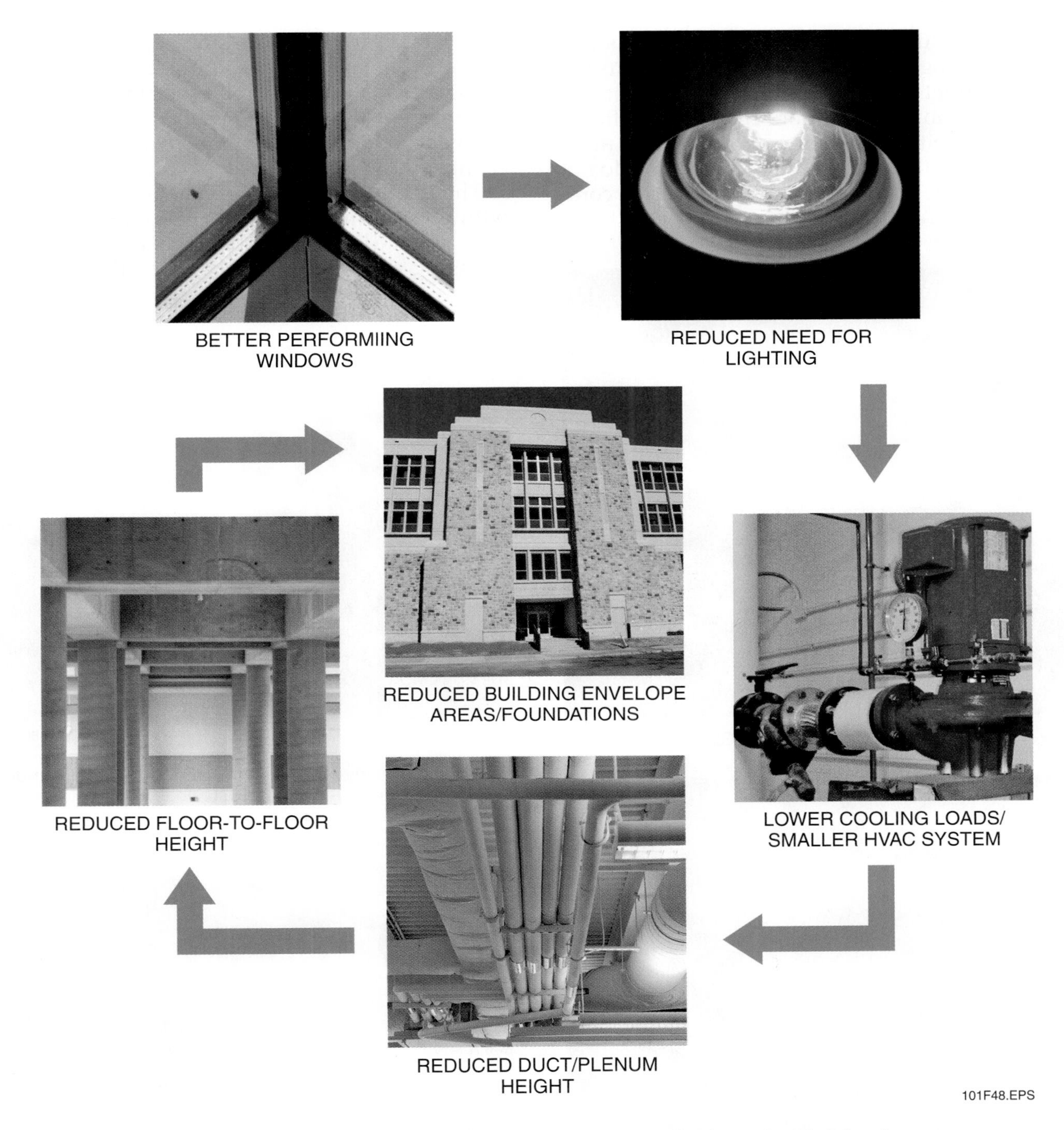

Figure 48 ◆ Integrated design means that investments in one system can be offset by savings in related systems.

2.7.4 Considering the Options

When considering what actions to take on a project, it's critical to make smart choices. Every action you take comes with a cost. Recognizing those costs and considering the benefits of your actions can help you choose actions that achieve the results you intend. Think about leverage points. Look for easy actions that can make a big difference.

One way to think about potential choices is to consider who has to act. Do they require action on the part of building professionals, laypersons, or both? As a building professional, you might not want to change your work practices. However, your knowledge of construction means that you have a much better chance of being successful than a building user getting involved with the operation of the building (*Figure 49*).

Another issue to consider is whether users will have to change their behaviors to achieve the desired effects. Some changes are completely transparent to users. Others require significant change in habits or procedures. Changes are more likely to have the desired outcomes if they do not require people to change their behavior or comfort level. For example, asking occupants to turn down the heat to an uncomfortable level is a change that is unlikely to be sustained over time (*Figure 50*). When possible, choose solutions that can get the job done without requiring users to change their habits.

The degree to which the change is compatible with existing infrastructure is important. Some changes, like lighting retrofits, can be as simple as changing a light bulb (*Figure 51*). Other changes

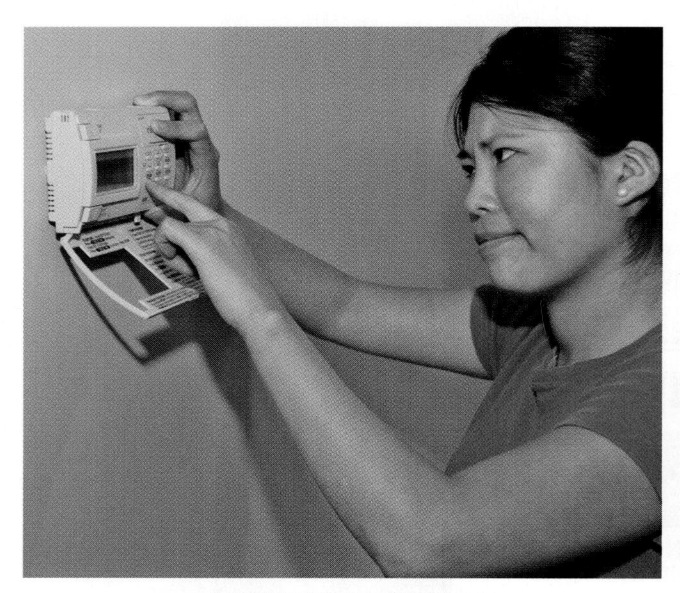

A. HOMEOWNER CHANGES

B. DESIGN CHANGES

101F49.EPS

Figure 49 ◆ Changes can be undertaken by professionals, laypersons, or both.

101F50.EPS

Figure 50 ◆ Changes that require users to sacrifice comfort are unlikely to be sustained.

101F51.EP

Figure 51 ◆ Changes can be simple if they require no modification of infrastructure.

may not be compatible at all with existing buildings or may not be possible given the skills and equipment of the construction crew. Look for easy changes that can be undertaken with your existing skills and tools wherever possible.

Finally, availability is an important consideration in determining how easy a change will be to implement. Imagine if you had to special order every piece of technology you needed to get a job done. It's much easier to make a change when you know you can easily obtain the materials and systems you need to get the job done. Many building material retail stores are beginning to stock green products (*Figure 52*). Be sure to ask about these products using the ratings and logos discussed in earlier parts of this module.

101F52.EPS

Figure 52 ◆ Some green products are easy to find on the shelves of your local store.

2.7.5 Counting All the Costs

Decisions about individual materials or systems are sometimes made on a first cost basis without considering cost from a life cycle perspective. This means that some green products seem more expensive than they really are. If savings in labor cost and construction schedule were taken into account, these materials could show an immediate cost advantage to owners. For example, one project manager estimated a savings of $3,000 per day due to shortening the construction schedule on a dormitory project. This dormitory was built using pre-engineered concrete panels, enabling the project to be built more quickly. In addition, the project saved money by avoiding the hotel expenses of dorm residents, who were able to move in sooner. Other factors that can make the cost of green building more reasonable include the following:

- Reduced costs of materials and waste disposal
- Reduced liability and environmental risk
- Happier occupants with increased productivity, reduced absenteeism, fewer building-related health problems, improved morale, and less employee turnover
- Reduced operational and disposal costs
- Reuse of facilities that otherwise would be disposed
- Preparedness for future regulations and requirements

Each of these benefits reflects a potential cost savings for owners. However, many of these types of costs are not typically counted as part of project costs. If these benefits can be realized, then green projects will have an economic advantage over conventional projects. Counting all the costs gives a truer picture of what green best practices will mean for the building over its whole life cycle.

Review Questions

1. The percentage of U.S. buildings that suffer from poor indoor air quality is about _____ percent.
   a. 15
   b. 20
   c. 33
   d. 50

2. The longest phase of the building life cycle is typically _____.
   a. planning
   b. design
   c. construction
   d. operation and maintenance

3. The parts of a building that are paved are called _____.
   a. softscape
   b. hardscape
   c. landscape
   d. parkscape

4. A way to minimize the urban heat island effect is to _____.
   a. pave with asphalt
   b. choose light-colored pavements
   c. choose dark-colored pavements
   d. coat paved areas with a sealer

5. Most of the water on the planet is _____.
   a. groundwater
   b. ice
   c. salt water
   d. fresh water

6. The first step in greening water use is to _____.
   a. identify alternative toilets
   b. eliminate unnecessary uses
   c. install low-flow toilets
   d. install a water meter

7. A common technique to reduce the actual flow of water in faucets is _____.
   a. turning off the water
   b. aeration
   c. heating
   d. cooling

8. A green alternative source of water is _____.
   a. groundwater
   b. rainwater harvesting
   c. melting ice
   d. salt water

9. Systems that capture water and redirect it for use in irrigation or toilet flushing are known as _____ systems.
   a. graywater
   b. blackwater
   c. brackish
   d. reclaim

10. A high-albedo roof _____.
    a. absorbs solar energy
    b. is expensive
    c. reflects solar energy
    d. reduces light pollution

11. Lighting retrofits to a building have a _____ payback period.
    a. slow
    b. rapid
    c. large
    d. negative

12. Green power generates energy from _____.
    a. nonrenewable resources
    b. coal
    c. renewable resources
    d. natural gas

Review Questions

13. Salvaged materials save the use of _____.
    a. raw materials
    b. labor
    c. equipment
    d. water

14. Material that is recycled from manufacturing processes is termed _____ waste.
    a. post-consumer
    b. consumer
    c. post-industrial
    d. pre-industrial

15. An example of a multi-function material is _____.
    a. structural insulated panels
    b. hardwood flooring
    c. concrete
    d. steel framing

16. A smart material changes in response to _____.
    a. funding
    b. the user's needs
    c. environmental conditions
    d. a schedule

17. Rapidly renewable materials are _____.
    a. recycled
    b. bio-based
    c. unsustainable
    d. expensive

18. Modern finishes, like paint, release _____ as they age.
    a. air
    b. VOCs
    c. carbon
    d. energy

19. Construction isolation measures to prevent building contamination should protect _____.
    a. storage areas
    b. HVAC intakes
    c. water supplies
    d. construction equipment

20. Passive survivability is the ability for a building to function when _____.
    a. the infrastructure goes down
    b. there is a labor strike
    c. there is a shortage of supplies
    d. the air is contaminated

3.0.0 ◆ TOOLS AND STRATEGIES

Choosing the best green practices to use on a project can be difficult, but there are tools to help. Green building rating systems help you evaluate how buildings affect the green environment. Green building rating systems are particularly important to owners and purchasers of buildings. They offer valuable information about the building's overall environmental performance. Green building rating systems provide a way to help the construction market do better in terms of meeting green project goals. Most green building rating systems include performance levels that must be met for the building to be certified. They may also provide guidelines that help project teams meet or exceed those levels.

There are two major types of green building rating systems: local and national. Local rating systems are typically developed by local builder associations. These systems take into account the local climate and building best practices for the area. Most local rating systems apply to residential buildings. There are over 25 different local or regional residential rating systems in existence. The first such program was established in Austin, Texas in 1991. Other local rating systems include **EarthCraft** homes in the southeastern U.S., the Built Green rating system in Colorado (*Figure 53*), and the California Green Builder program for production builders in California.

101F53.EPS

Figure 53 ◆ The Built Green program certified this home outside Denver, Colorado.

The second type of rating system is national. At the national level, there are both residential and commercial programs. Many national rating systems used in other countries are modeled after the U.S. Green Building Council's **Leadership in Energy and Environmental Design (LEED)®** rating system (see www.leedbuilding.org), developed in the early 1990s. The LEED rating system was based on a similar system called the Building Research Establishment Environmental Assessment Method (BREEAM), which was developed earlier in the United Kingdom.

National residential rating systems include LEED for Homes and the National Association of Home Builders' National Green Building Standard (www.nahbrc.org). Commercial rating systems at the national level include various versions of the LEED rating system and the Green Globes rating system (www.greenglobes.com).

There are also rating systems that apply at the international level. The Sustainable Building Challenge managed by the International Initiative for a Sustainable Built Environment (www.iisbe.org) is one example. International programs are much less common than national rating systems.

The LEED Green Building Rating System is the predominant standard in the U.S. for rating commercial buildings (*Figure 54*). LEED is a reference standard for government agency buildings at the federal, state, and local levels. It has also been adopted by multiple owners in the private sector as well. The U.S. General Services Administration

101F54.EPS

Figure 54 ◆ The LEED Green Building Rating System ensures that buildings are constructed to specific environmental standards.

has a policy requiring all new buildings to meet at least the Silver level of LEED certification. Other federal agencies have followed suit. Many states also require LEED certification for public buildings at various levels. Even cities have adopted LEED as a standard, including the cities of New York, Atlanta, Seattle, and others. By 2010, the U.S. Green Building Council (USGBC) estimates that approximately 10 percent of all new construction projects in the United States will pursue LEED certification.

3.1.0 LEED Green Building Rating System

The LEED rating system applies to a wide variety of project types. It is designed to be applicable in different climates and contexts throughout the U.S. It is characterized in terms of its structure, seven major categories of credits, and four levels of certification. A complete list of the LEED prerequisites, credits, and point values for new construction (LEED-NC Version 3/2009) is provided in the *Appendix*.

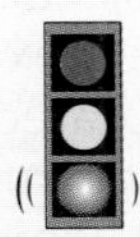

NOTE

This module focuses primarily on the LEED for New Construction (LEED-NC) rating system, in Version 3 at the time of this writing. Other rating systems differ by project type and may vary in specific credit requirements.

3.1.1 Structure of the LEED Rating System

LEED for New Construction (LEED-NC) was the first of all the LEED rating systems to be developed. Each of the subsequent systems was modeled after the same structure. The LEED rating system consists of a series of performance goals and requirements in seven primary categories:

- *Sustainable Sites (SS)* – This category covers issues related to the location of the project site, impacts to the site during construction, site amenities, and impacts resulting from building operations.
- *Water Efficiency (WE)* – This category addresses water consumption and wastewater generation by the building in operation.
- *Energy and Atmosphere (EA)* – This category covers all aspects of the building's energy performance, energy source(s), and atmospheric impacts.
- *Materials and Resources (MR)* – This category pertains to the sources and types of materials used on the project, the amount of waste generated, and the degree to which the project makes use of existing buildings.
- *Indoor Environmental Quality (EQ)* – This category covers aspects of the building's indoor environment, ranging from ventilation to air quality to daylight and views.
- *Innovation in Design (ID)* – This category rewards the project for going beyond the minimum credit requirements and for using a LEED Accredited Professional.
- *Regional Priority (RP)* – This category rewards projects for addressing local environmental priorities (identified by ZIP code at www.usgbc.org).

Each category consists of a series of credits that define points that can be earned by a project. Five out of the seven categories also have prerequisites that the project must meet to be considered for certification. A project must meet all prerequisites

Seattle Focuses on Saving Power

Seattle, Washington, has written LEED certification into its City Plan. In addition to providing city staff to assist builders in creating and building green projects, they have legislated that all city-funded projects and renovations with over 5,000 square feet of occupied space must achieve a LEED Silver rating.

101SA15.EPS

in all categories in order to pursue certification. It also must obtain at least two energy performance credits to exceed the energy code requirements. Regional priority credits may apply as bonus credits to any of the other credits, depending on local concerns. For example, a system that pays special attention to water efficiency in a desert environment might be given an extra RP credit in addition to a WE credit.

3.1.2 Types of Rating Systems and Levels of Certification

The LEED system was initially developed to apply to new commercial construction. As the system grew in popularity, it became apparent that different types of projects would require different criteria to be properly rated. The original rating system was then customized through the development of application guides for specific project types such as hotels and dormitories. These types of projects have characteristics that require special interpretation of LEED credit requirements. However, they can still use the basic LEED-NC structure. Separate versions of the rating system were also developed for the following types of projects:

- Existing Buildings (LEED-EB) for buildings not rated during construction that have already been occupied
- Commercial Interiors (LEED-CI) for individual tenant spaces in commercial buildings
- Core and Shell (LEED-CS) for the site and base building for buildings that lease tenant space
- Homes (LEED-H) for residential construction
- Neighborhood Development (LEED-ND) for residential and mixed-use developments of various types

LEED-CS and LEED-CI are complementary products that reinforce each other. For example, tenant spaces seeking certification under LEED-CI can obtain points for locating in buildings that have received LEED-CS certification. The same applies to LEED-H and LEED-ND. Homes seeking certification under LEED-H can receive extra points if the neighborhood in which they are developed has received LEED-ND certification. Likewise, additional points can be earned under LEED-ND if the buildings in the development are also certified under an appropriate rating system such as LEED-H or LEED-NC.

There are 110 possible points under the LEED-NC rating system, and eight prerequisites. Projects must meet all eight prerequisites to pursue certification. Certification can be achieved at four different levels:

- Certified: 40 to 49 points
- Silver: 50 to 59 points
- Gold: 60 to 79 points
- Platinum: 80 points and above

Platinum is the most difficult level to reach. Currently, only about 120 projects have achieved this level of certification. Each type of rating system has different point thresholds since the credits and points differ slightly from system to system. However, all LEED rating systems award certification at the four listed levels.

3.1.3 Certification Process

The process of certifying a building under LEED has several steps. Each of these steps is undertaken using the LEED Online documentation system. The first step in the process is to register the project with the USGBC to declare intent to pursue certification. This allows the project team to access USGBC databases as well as set up an online workspace to manage project documentation. Registering the project requires paying a flat fee to the USGBC that is the same for all projects.

After the project has been registered, the next step is to document green features incorporated into the project design that meet credit requirements. Typically, the project team will kick off this step by having a meeting to review the LEED checklist (*Figure 55*) and decide which points and credits to pursue. The process of documentation to prove compliance with LEED credit requirements continues throughout construction and possibly for the first year of occupancy depending on the credits pursued.

The project team can submit documentation to the USGBC for review at two points in time: at the end of the design process, and after construction is complete. The project team may also opt to submit everything at once when the project is complete. All documentation compiled up to the point of submittal will have been assembled by the project team using LEED Online. Submittal of documentation for review involves paying a review fee based on project size. This prompts the USGBC to conduct the review. If the team submits a design review, USGBC will evaluate points under the rating system that can be measured at the end of the design process. It will not evaluate credits that require documentation during the construction phase. Design phase review is useful for the project team to get an idea of how many points they are likely to obtain in the project. It provides a basis for deciding how hard to work for additional points during construction.

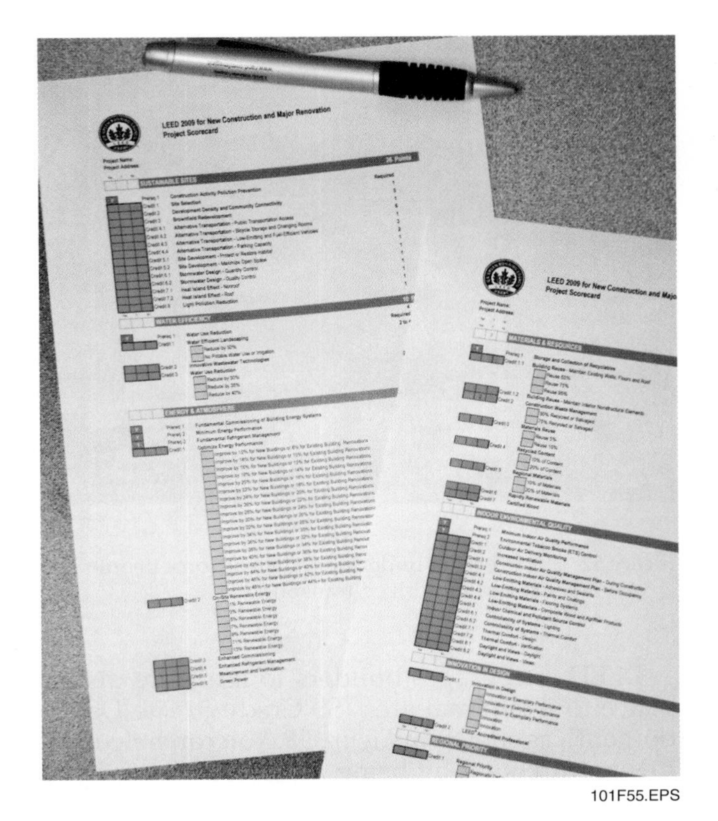
LEED 2009 for New Construction and Major Renovation
Project Scorecard

101F55.EPS

Figure 55 ◆ LEED project worksheet.

At the conclusion of the project, final documentation is assembled online. A fee is paid, and the package is reviewed by the USGBC. Upon review of the full project documentation, USGBC may elect to request additional clarification as part of a point audit. The Council makes a final determination as to which points should be awarded and determines a level of certification for the project. The project team has the right to appeal any credits declined in the review by paying an additional fee and providing additional documentation. The ruling of the USGBC following appeals is final.

3.2.0 Goals of the LEED Green Building Rating System

The basic concepts of the LEED rating system can be captured in the form of eight simple goals:

1. Pick a good site for your project.
2. Protect and restore your surroundings.
3. Provide amenities that promote sustainable behavior.
4. Use ecologically friendly resources.
5. Find better waste sinks.
6. Protect indoor environmental quality.
7. Check systems to be sure they work right.
8. Look for better ways to do things.

DID YOU KNOW?

Accreditation vs. Certification

A common mistake when discussing LEED projects is to confuse accreditation and certification. Accreditation is a process that is applied to building professionals who manage the process of LEED certification for projects. Becoming a LEED Accredited Professional requires you to pass the LEED Accreditation Exam (more information is available at www.gbci.org, the website of the Green Building Certification Institute that manages and administers the LEED exam). Certification, on the other hand, applies to buildings. Buildings are certified under the LEED rating system and are awarded certification at the Certified, Silver, Gold, or Platinum levels. Keeping these terms straight will help your credibility with respect to the LEED rating system. Remember, people are accredited and buildings are certified.

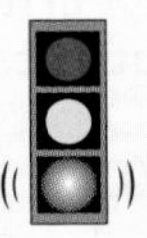

NOTE

The details of LEED credit requirements are beyond the scope of this module. If you're interested in learning more about specific LEED credit requirements, you can download the free LEED standards for any of the current rating systems at www.usgbc.org.

3.2.1 Selecting the Site

The first goal of the LEED rating system is to help you pick a good site for your project. LEED rewards projects that avoid developing vulnerable or ecologically valuable sites. These sites include wetlands, habitats of threatened or endangered species, or flood plains (SS Credit 1: Site Selection). The rating system also awards credits for developing in areas that are already developed to a certain density, as shown in *Figure 56*. These sites are preferred over undeveloped sites (SS Credit 2: Development Density and Community Connectivity) that can require the use of motorized transportation, as shown in *Figure 57*. LEED also encourages projects to locate on sites that are brownfields (SS Credit 3: Brownfield Redevelopment). Brownfields are sites that have real or perceived environmental contamination, such as old gas stations, industrial sites, and dry cleaners.

101F56.EPS

Figure 56 ◆ Sites close to amenities reduce the need to travel by car.

101F57.EPS

Figure 57 ◆ Projects in undeveloped areas force people to use motorized transport.

3.2.2 Protecting and Restoring the Site

The second major goal of the LEED system is avoiding damage to the site and/or building during construction (EQ Credit 3: Construction Indoor Air Quality Management Plan). LEED also encourages ecological site restoration at the end of construction. The construction team plays a critical role in achieving the credits and prerequisites under this goal. A prerequisite for all LEED projects is to meet all requirements of the 2003 U.S. EPA Construction General Permit, which includes measures to protect soil and water streams, and to mitigate dust and noise. All projects over one acre in size must comply with these requirements under federal law.

LEED encourages builders to limit site disturbance to a minimal area (SS Credit 5: Site Development), as shown in *Figure 58*. You can help meet this requirement by being aware of how far you can go outside your work boundary. If you are working on a LEED project, ask your supervisor to mark the boundaries and be sure you stay within them.

The LEED rating system rewards buildings that treat and/or retain stormwater on site to prevent contamination of local waterways (SS Credit 6: Stormwater Design). SS Prerequisite 1 (Construction Activity Pollution Prevention) requires projects to use erosion controls to prevent stormwater runoff, **sedimentation**, and dust from leaving the

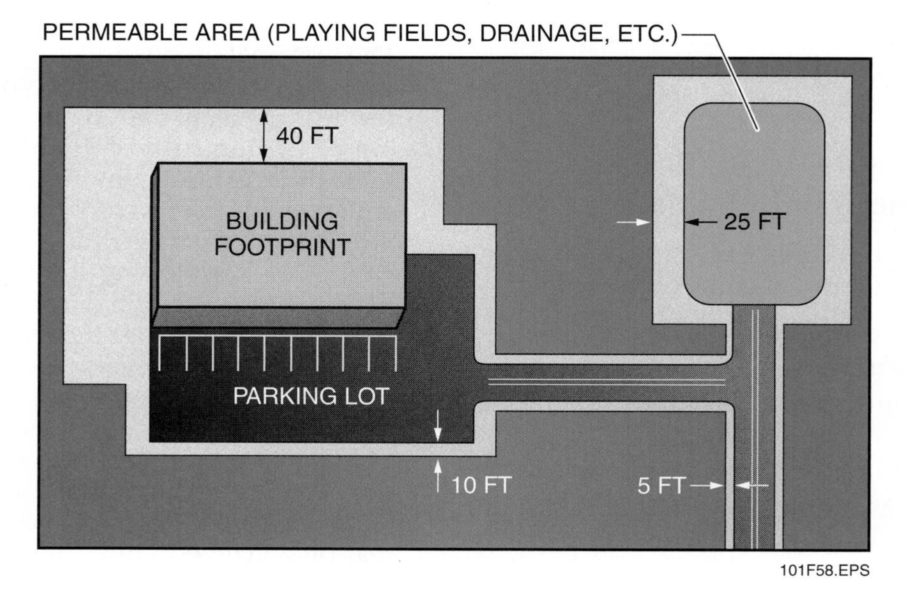

101F58.EPS

Figure 58 ◆ Site disturbance must stay within these limits to meet LEED requirements.

site (*Figure 59*). Another consideration is managing excavated soil (*Figure 60*). The EPA General Permit Requirements include stabilizing spoil piles.

Install protective fencing when working near trees. Place the fencing at or beyond the drip line of the tree's outermost branches to protect the root system (*Figure 61*). Notify your supervisor if you see unprotected trees.

Projects that use light-colored roofs and parking areas are recognized for reducing urban heat islands (SS Credit 7: Heat Island Effect). Urban heat islands are developed areas where ambient temperatures are higher than the surrounding area. They are caused by dark-colored pavements and buildings that absorb solar energy (*Figure 62*).

101F59.EPS

Figure 59 ◆ Sedimentation fencing in place to manage stormwater runoff.

101F60.EPS

Figure 60 ◆ Grass seeding and straw mulch are one way to stabilize excavated soil.

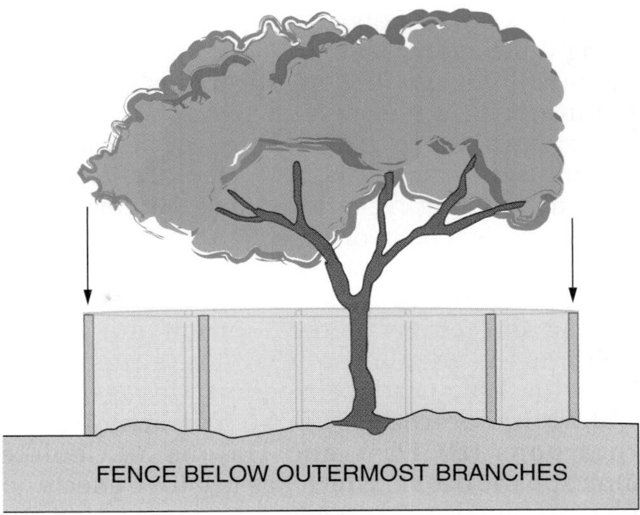

101F61.EPS

Figure 61 ◆ Effective tree protection fencing requires the tree to be protected as far out as the drip line to avoid crushing tree roots.

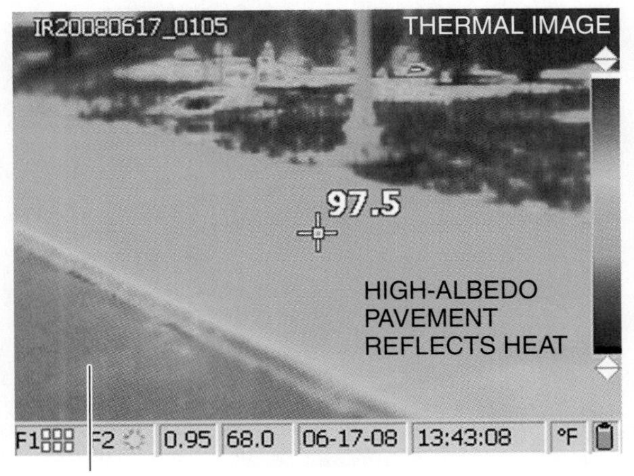

101F62.EPS

Figure 62 ◆ Urban heat islands are worsened by dark pavements.

LEED also acknowledges projects that avoid using unnecessary levels of lighting (SS Credit 8: Light Pollution Reduction) due to unshaded light fixtures, such as the one shown in *Figure 63.* Light pollution disturbs endangered species such as loggerhead turtles. It is suspected of disrupting sleep and mating cycles in animals, and may have adverse effects on people.

Projects are encouraged to reduce or eliminate the use of ozone-depleting chemicals in air conditioning and refrigeration equipment and as propellants in insulation (EA Prerequisite 3/ Credit 4: Enhanced Refrigerant Management). **Chlorofluorocarbons** (**CFCs**), **hydrochlorofluorocarbons (HCFCs)**, and **Halons** all deplete atmospheric ozone, which has negative effects on plants, animals, and people.

3.2.3 Providing Amenities that Promote Sustainable Behavior

The next major goal of the LEED rating system encourages projects to include amenities that promote sustainable behavior. This includes providing amenities that encourage people to use low-impact transportation such as bicycling or carpooling (SS Credit 4: Alternative Transportation), as shown in *Figure 64.* Water and energy performance credits require the building to include efficient fixtures that save water and energy during its life cycle (WE Prerequisite 1: Water Use Reduction and EA Prerequisite 2/Credit 1: Optimize Energy Performance).

Recycling facilities (*Figure 65*) encourage occupants to sort their solid waste and reduce the amount of trash that goes to landfills (MR Prerequisite 1: Storage and Collection of Recyclables).

Providing smoking facilities as required by EQ Prerequisite 2: Environmental Tobacco Smoke (ETS) Control helps protect nonsmokers from environmental tobacco smoke. When adequate smoking facilities are not provided, smokers tend to huddle near the door, which may cause discomfort for those entering and leaving the building. Many hospitals, schools, and government buildings prohibit smoking within a certain distance of building entrances.

101F64.EPS

Figure 64 ◆ Bike racks, reserved parking for carpools, and preferred parking for motorcycles and high-efficiency vehicles reward green behavior.

101F63.EPS

Figure 63 ◆ Many outdoor fixtures contribute to light pollution.

101F65.EPS

Figure 65 ◆ Recycling facilities encourage users to sort their waste.

3.2.4 Using Ecologically Friendly Resources

The fourth goal encompassed by LEED is to use the least expensive resources you can find from an ecological standpoint. This can include using graywater or rainwater for landscaping (WE Credit 1: Water Efficient Landscaping) or toilet flushing (WE Credit 2: Innovative Wastewater Technologies), such as the under-sink graywater system shown in *Figure 66*. It also can involve the use of materials that are part of an existing building (MR Credit 1: Building Reuse), salvaged materials (MR Credit 3: Materials Reuse), from a recycled source (MR Credit 4: Recycled Content), or from a rapidly renewable source (MR Credit 6: Rapidly Renewable Materials). Finding better sources for electrical energy is encouraged via installation of on-site power generation (EA Credit 2: On-Site Renewable Energy). You can also get credit for contracting for green power from a third party provider (EA Credit 6: Green Power).

The use of local or regional materials is encouraged (MR Credit 5: Regional Materials). *Figure 67* shows an example of the 500-mile radius that defines a regional material according to LEED.

As a craft worker, you may be required to procure green materials and systems to meet the project specifications (*Figure 68*). When procuring these materials, carefully review the specifications to be sure you understand all product requirements. Keep cut sheets or other documentation to support compliance. These materials will help when preparing the project documentation.

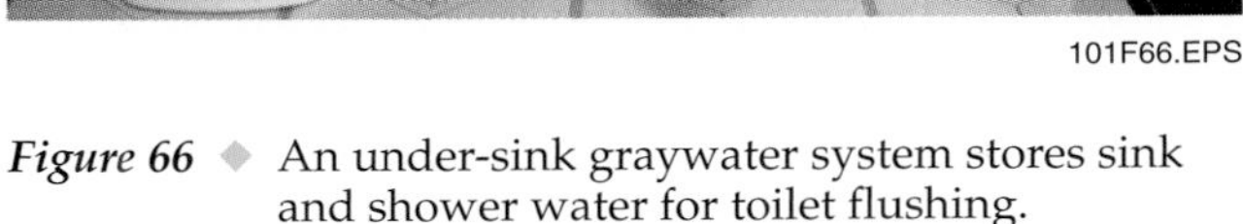

101F66.EPS

Figure 66 ◆ An under-sink graywater system stores sink and shower water for toilet flushing.

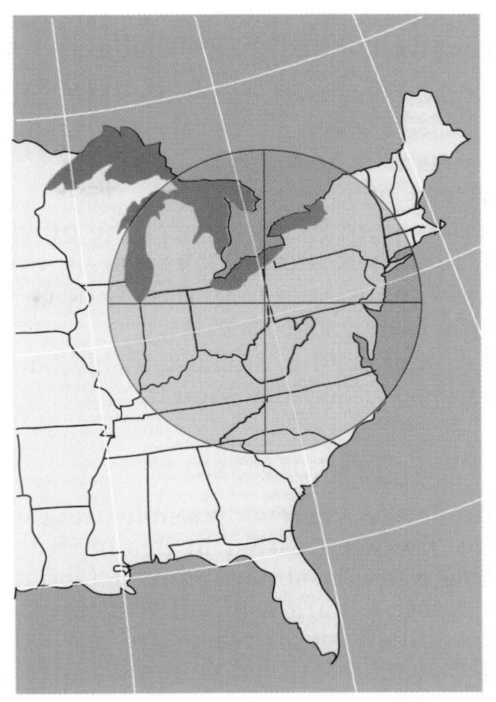

101F67.EPS

Figure 67 ◆ A regional material comes from within 500 miles of the job site.

A. Paper fiber panel core
B. Laminated strand lumber (LSL)
C. Engineered wood I-joist
D. Abundant American hardwoods
E. Laminated veneer lumber (LVL)
F. Adobe block with straw
G. Straw-core SIP with oriented strand board (OSB) skin
H. Parallel strand lumber (PSL)

101F68.EPS

Figure 68 ◆ Bio-based materials are an important resource for green building.

Better sources for construction materials that are encouraged under LEED include:

- Salvaged or reused materials (MR Credit 3: Materials Reuse)
- Materials with recycled content (MR Credit 4: Recycled Content)
- Materials from locations near the project site (MR Credit 5: Regional Materials)
- Rapidly renewable materials (MR Credit 6: Rapidly Renewable Materials)
- Wood products that are sustainably harvested (MR Credit 7: Certified Wood)

3.2.5 Finding Better Waste Sinks

Minimize waste wherever possible using the best practices discussed earlier in this module. With remaining waste, be sure you understand the policy for **waste separation** and recycling. Better sinks for construction waste are rewarded by LEED (MR Credit 2: Construction Waste Management). Construction projects can meet this credit requirement by sending mixed waste to an off-site separation facility (*Figure 69*) or through on-site separation of construction and demolition materials for recycling (*Figure 70*).

If on-site recycling separation is part of the project, be sure to put waste in the correct dumpster (*Figure 71*). If you don't see a place to recycle, ask for one. Do not contaminate dumpsters with other waste. In particular, keep treated wood separate from other materials. Also, keep anything with food on it out of cardboard dumpsters. Cleaner dumpster loads are more likely to be accepted for recycling.

For excess materials or salvaged goods on a project, ask your supervisor to arrange a place for temporary storage. Consider donating unused materials to a Habitat for Humanity ReStore.

101F70.EPS

Figure 70 ◆ In other areas, on-site sorting is required for recycling.

101F69.EPS

Figure 69 ◆ In some areas, mixed construction waste can be taken off site and sorted for recycling.

101F71.EPS

Figure 71 ◆ On-site waste separation can earn up to two LEED credits plus one Innovation Credit.

If there is a ReStore near you, take the time to visit it. You may find products that you want to buy.

3.2.6 Protecting Indoor Environmental Quality

The sixth major goal of the LEED rating system focuses on designing buildings to be amenable to human needs. LEED strives to make buildings comfortable in terms of the following:

- Achieving minimum levels of indoor air quality through good building design (EQ Prerequisite 1: Minimum Indoor Air Quality Performance)
- Keeping nonsmokers from being exposed to environmental tobacco smoke [EQ Prerequisite 2: Environmental Tobacco Smoke (ETS) Control]
- Making sure spaces receive adequate ventilation (EQ Credit 1: Outdoor Air Delivery Monitoring and EQ Credit 2: Increased Ventilation)
- Avoiding the use of materials that emit high levels of VOCs (EQ Credit 4: Low-Emitting Materials). *Figure 72* shows a product label indicating VOC content.
- Separating areas of the building that are likely to contain polluting activities from other areas (EQ Credit 5: Indoor Chemical and Pollutant Source Control)
- Providing users with the ability to control conditions in their own spaces (EQ Credit 6: Controllability of Systems)
- Achieving thermal comfort for at least 80 percent of building occupants (EQ Credit 7: Thermal Comfort)
- Providing users with access to natural daylight and views as much as possible (EQ Credit 8: Daylight and Views)

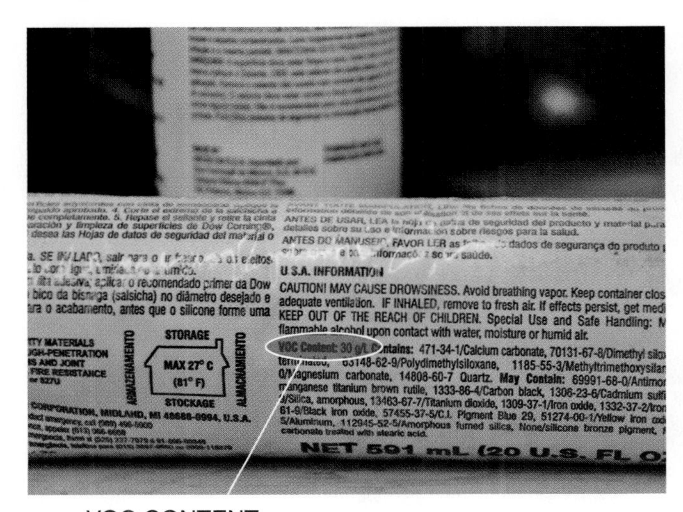

Figure 72 ◆ Look for levels of VOCs on product labels and buy products with low levels.

If you work inside the building, make sure you understand how your activities affect indoor air quality. Consider whether your activities will produce airborne dust or offgas VOCs. Take measures to protect the building from these activities (*Figure 73*).

To create buildings that are free of problems for future occupants, focus on activities inside the building. Using low-emitting materials can help to ensure good indoor air quality (*Figure 74*).

This item also covers areas such as chemical mixing and storage areas and copy rooms (*Figure 75*). Copiers and printers produce ozone during operation. Ozone causes respiratory irritation.

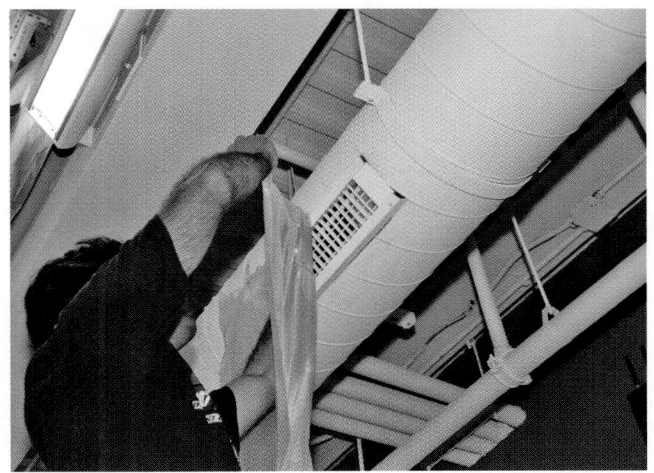

Figure 73 ◆ Containing construction contaminants is critical to ensure indoor air quality.

Figure 74 ◆ Low-emitting materials can help to ensure good indoor air quality.

101F75.EPS

Figure 75 ◆ Photocopiers are a common source of indoor air pollution in office environments.

Prevent moisture damage by storing materials in a protected staging area. Never install wet materials. Installing wet or damp materials can create mold problems in the building before it is even complete.

The last credit, Daylight and Views, is included because of growing evidence that these factors provide numerous benefits to occupants (*Figure 76*).

101F76.EPS

Figure 76 ◆ Daylight and views are correlated with many positive outcomes.

Exposure to natural daylight has been correlated with effects ranging from better test scores to improved dental health. Being able to look out the window has also been linked with increased productivity in offices and faster healing in hospitals. LEED encourages the use of daylighting to increase human comfort as well as reduce the need to rely on electric lighting.

3.2.7 Performing System Checks

LEED puts special emphasis on checking the design, construction, and initial operation of a building to be sure it meets the design intent. It

High-Performance Building Envelopes

Rinker Hall on the University of Florida's Gainesville campus earned LEED Gold certification for its energy-efficient design. The architect was challenged to design a building that would fit in with the other brick buildings on campus, but not trap heat. The compromise was a freestanding masonry shade wall that serves as a second skin and helps match the design of the building to the site. The high-performance building envelope reflects heat and includes an air-infiltration barrier, a thermal break, and high-performance glass.

101SA16.EPS

requires a process called commissioning (EA Prerequisite 1: Fundamental Commissioning of Building Systems/Credit 3: Enhanced Commissioning). This process involves a third party in design review, testing and balancing of systems, and system turnover (*Figure 77*).

LEED also rewards projects that take measures to verify and maintain performance of systems (EA Credit 5: Measurement and Verification). This can be done by installing control systems that measure performance. These systems also adjust the building's mechanical and electrical systems to optimize performance. Points are awarded for using CO_2 monitors to control building ventilation (EQ Credit 1: Outdoor Air Delivery Monitoring) and designing the building to keep occupants comfortable (EQ Credit 7: Thermal Comfort). This last credit also offers a point for doing a post-occupancy evaluation one year after the building has been occupied. The purpose of this survey is to ensure that building users are still comfortable after the building has been broken in.

3.2.8 Seeking New Methods

LEED rewards projects for finding new ways to make projects green that are not currently included in the rating system (ID Credit 1: Innovation in Design). Project teams can propose new credits for future versions of LEED based on best practices they have achieved. One example of a successful Innovation Credit is green housekeeping using nontoxic, bio-based chemicals (*Figure 78*). Projects can also receive Innovation Credits for greatly exceeding credit thresholds in the areas of water efficiency, energy performance, materials use, and waste diversion. Finally, projects can be awarded a point for involving a LEED Accredited Professional (LEED AP) as part of the project team (ID Credit 2: LEED Accredited Professional), or by earning Green Advantage® contractor certification. Ask your supervisor about additional training for green projects such as the Green Advantage® training program (*Figure 79*) or the LEED AP exam.

Be aware of the project's goals for LEED certification and speak up if you have ideas for improvement. On the job, you may be impacted by various Innovation Credits such as green housekeeping. Another possible impact may happen if your project is pursuing Innovation Credits for education. You may see students on the job site who are observing or recording your work as part of LEED documentation of the project. Feel free to ask questions. If you see students observing, it is important to show them your typical work practices so they can accurately capture what happens on the job site.

101F77.EPS

Figure 77 ◆ Building commissioning provides a chance to check that all building systems work well together.

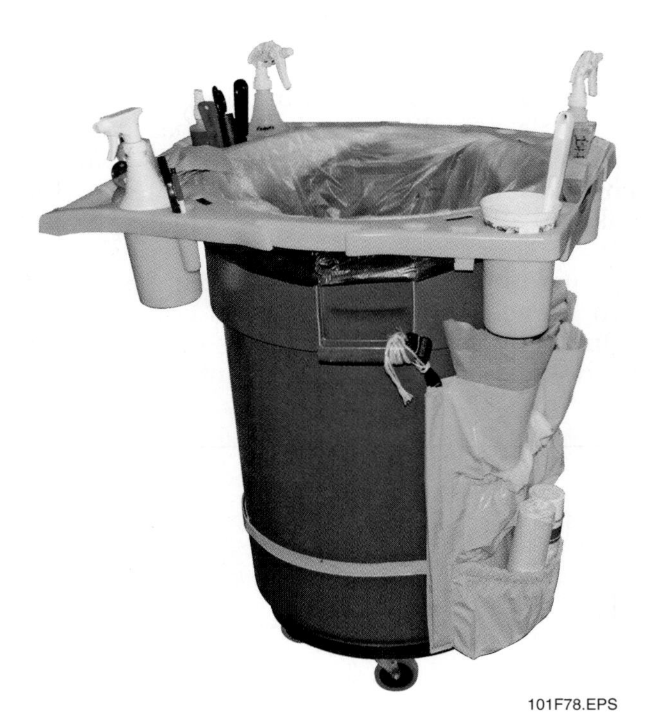

101F78.EPS

Figure 78 ◆ Green housekeeping programs can earn your project a LEED Innovation Credit.

101F79.EPS

Figure 79 ◆ Green Advantage® is an environmental certification for contractors that can make your commercial project eligible for a LEED Innovation Credit.

3.3.0 LEED Documentation

Managing documentation for a LEED project can be challenging. The LEED AP handling the documentation of your project will be grateful if your materials and activities are well documented. Be sure you understand the types of information you will be required to provide, and ask questions if needed.

Figure 80 shows a typical form used to track information to perform LEED calculations. As you can see, each material must have cost information in order to be included in the spreadsheet. For recycled content, you need the percent by weight of both post-consumer and pre-consumer recycled content for all parts of the material. As you might imagine, this can get quite complicated for a product such as carpeting, which includes several different materials.

Another challenging credit to calculate is MR Credit 5: Regional Materials. This credit requires that you know the origin of the material and its components. A material may be counted for this credit only if it and all its components originated within a 500-mile radius of the site.

Materials Credits are calculated based on percent of total materials cost. Therefore, it makes sense to focus your attention on materials that make up a large share of the total material costs. You can sometimes achieve LEED credits just by focusing on steel and concrete. Steel has significant recycled content, as does concrete if it includes fly ash. These materials often make up the most expensive parts of the entire project.

DID YOU KNOW?

When in Doubt, Document

If you ever have questions about whether or not you should document something, always err on the side of caution. Gathering documentation after the fact is nearly impossible. Develop a documentation plan with your team and create a system for collecting and storing receipts, bills of lading, cut sheets, and other documentation relevant for LEED credits. Coordinate with other members of the project team to make sure you know what to keep and what to do with it.

3.4.0 Common Pitfalls During Construction

With all the ways you can contribute to a LEED project, there are also plenty of pitfalls to avoid. Three of the most serious pitfalls include poor execution of credit requirements, poor documentation, and lack of coordination with other trades.

3.4.1 Poor Execution of Credit Requirements

Even if you know exactly what is necessary to achieve a LEED credit, it is not always easy to meet those requirements. Some credits can be particularly difficult to achieve. For example, EQ

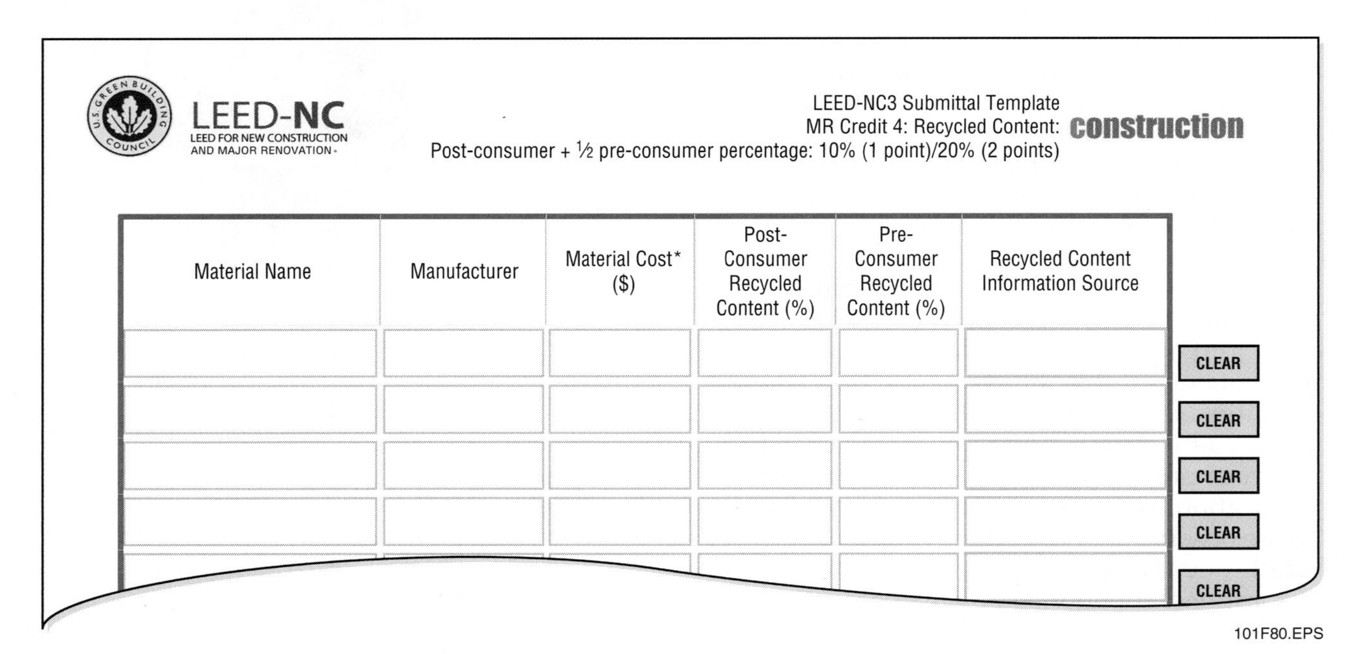

LEED-NC
LEED FOR NEW CONSTRUCTION AND MAJOR RENOVATION

LEED-NC3 Submittal Template
MR Credit 4: Recycled Content: **construction**
Post-consumer + ½ pre-consumer percentage: 10% (1 point)/20% (2 points)

Material Name	Manufacturer	Material Cost* ($)	Post-Consumer Recycled Content (%)	Pre-Consumer Recycled Content (%)	Recycled Content Information Source	
						CLEAR
						CLEAR
						CLEAR
						CLEAR
						CLEAR

101F80.EPS

Figure 80 ◆ Electronic form for tracking LEED Materials Credits.

Credit 3: Construction Indoor Air Quality Management Plan requires you to protect the indoor air quality of the building by taking various measures to isolate your work area and keep it clean. Failure to maintain these measures, even if you set them up correctly, can cause the project's goals not to be met.

Another potential problem area is SS Prerequisite 1: Construction Activity Pollution Prevention. As a prerequisite, these requirements must be met or the project's LEED certification can be denied. If you are responsible for maintaining construction pollution prevention for erosion and dust control, be sure you understand exactly what needs to be done. Check these systems regularly and repair them if they become damaged or dislodged. Even if these systems are not directly your responsibility, inform your supervisor of any problems so they can be corrected. Similarly, be sure you understand project requirements for SS Credit 5: Site Development. Ask questions if you are unclear. Bring any problems to the attention of your supervisor or crew leader.

MR Credit 2: Construction Waste Management is also a common opportunity for things to go wrong. Be sure you understand what the project's recycling goals are, and how they will be achieved. Ask questions in weekly or daily toolbox meetings, especially if you notice that others are not recycling properly. If you see contamination in any of the recycling bins, mention it to your supervisor. Talk to your crewmates to be sure they understand the importance of correct separation for recycling.

Procuring green materials for a project can also be a challenge. The specifications for a green project have been carefully developed to meet project green goals and should be followed whenever procuring materials. Ask vendors for documentation to support the products you buy.

3.4.2 Poor Documentation

The second major pitfall that can impede your project's LEED certification is sloppy documentation on the part of members of the project team. The amount of information needed to certify a project is extensive. Not only must you know the properties of the products you use, you also must report their cost. Therefore, keeping track of information as it occurs can help you successfully provide the information the project team needs.

One credit that requires careful documentation is MR Credit 2: Construction Waste Management. For this credit, you must show where all of the waste for your project has gone. This is necessary to calculate the percent of waste that was diverted from going to a landfill. To do this, you must save landfill tipping receipts and bills of lading from recycling companies. Set up a process for tracking this information.

A second credit requiring documentation is EQ Credit 4: Low-Emitting Materials. This credit requires you to show that carpets, paints, sealants, adhesives, and composite wood and agrifiber products meet VOC emission requirements. The easiest way to do this is to ensure that each product meets requirements before you bring it on the job site. Then, keep an inventory of products you have used and their labeled levels of VOCs.

What's Wrong with This Picture?

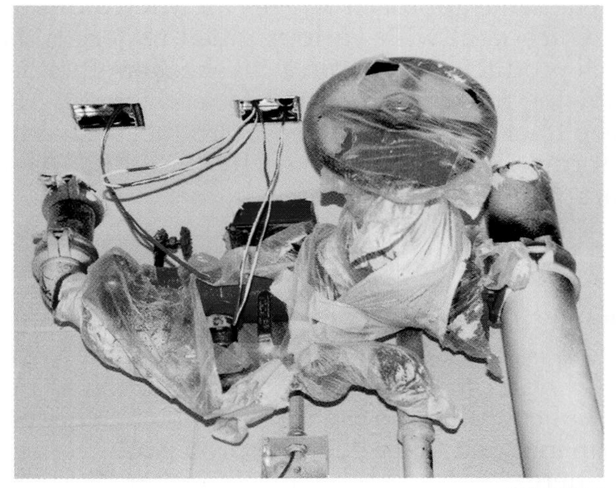

101SA17.EPS

What's Wrong with This Picture?

101SA18.EPS

For carpets and paints that are certified via a label, be sure to note the label when checking for compliance (*Figure 81*).

The same approach applies for MR Credits 3 to 7, which deal with various green materials. For each product used, retain a cut sheet or other data showing relevant properties for each credit. Also, keep track of the costs of each product so that this information can be tracked by the LEED AP as part of LEED calculations.

3.4.3 Lack of Coordination with Other Trades

The third major pitfall to achieving LEED certification is a lack of coordination with other trades. In addition to putting LEED certification at risk, lack of coordination can result in poor energy performance and other building problems.

Energy performance typically suffers the worst from poor coordination. All the cracks and gaps between materials installed by different trades can allow conditioned air to leak from the building. These leaks can cause significant energy loss. Careful construction follow-up can seal these leaks. However, this is often overlooked.

Coordination problems can also contribute to **thermal bridging**. Thermal bridging occurs where materials of higher conductivity (like metal and wood) are allowed to pass completely through the wall without a thermal break (*Figure 82*). These materials act as thermal pipelines and can result in cold spots in the wall or ceiling surface where moisture can condense. Over time, these moist areas serve as a breeding ground for mold.

101F81.EPS

Figure 81 ◆ Verify that products meet specifications by checking VOC levels on labels.

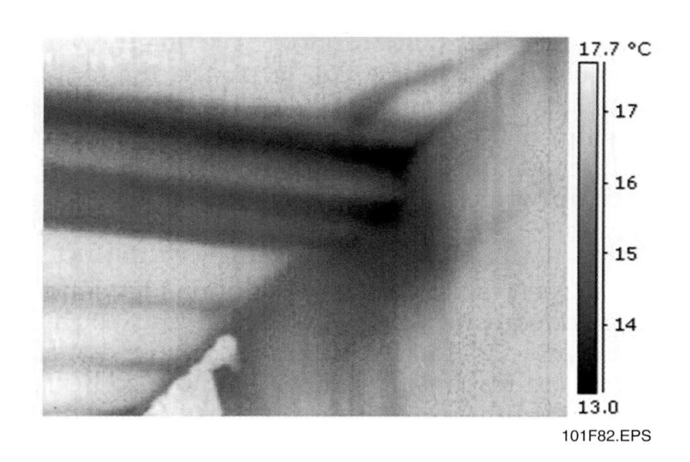

101F82.EPS

Figure 82 ◆ Thermal bridging caused by a metal beam.

Lack of coordination can also affect structural integrity. Conflicts inevitably occur when multiple trades must fit their systems into the same framing cavities. Improper notching can occur when plumbing or electrical trades install piping or conduit through structural members. The International Building Code limits the depth and location of cuts in all structural members (*Figure 83*). This is to ensure that structural integrity is preserved. Careful coordination of the trades in these areas prevents the need for rework. It can also reduce the risk of failure. Coordination is important in areas where considerable piping occurs.

What's Wrong with This Picture?

101SA19.EPS

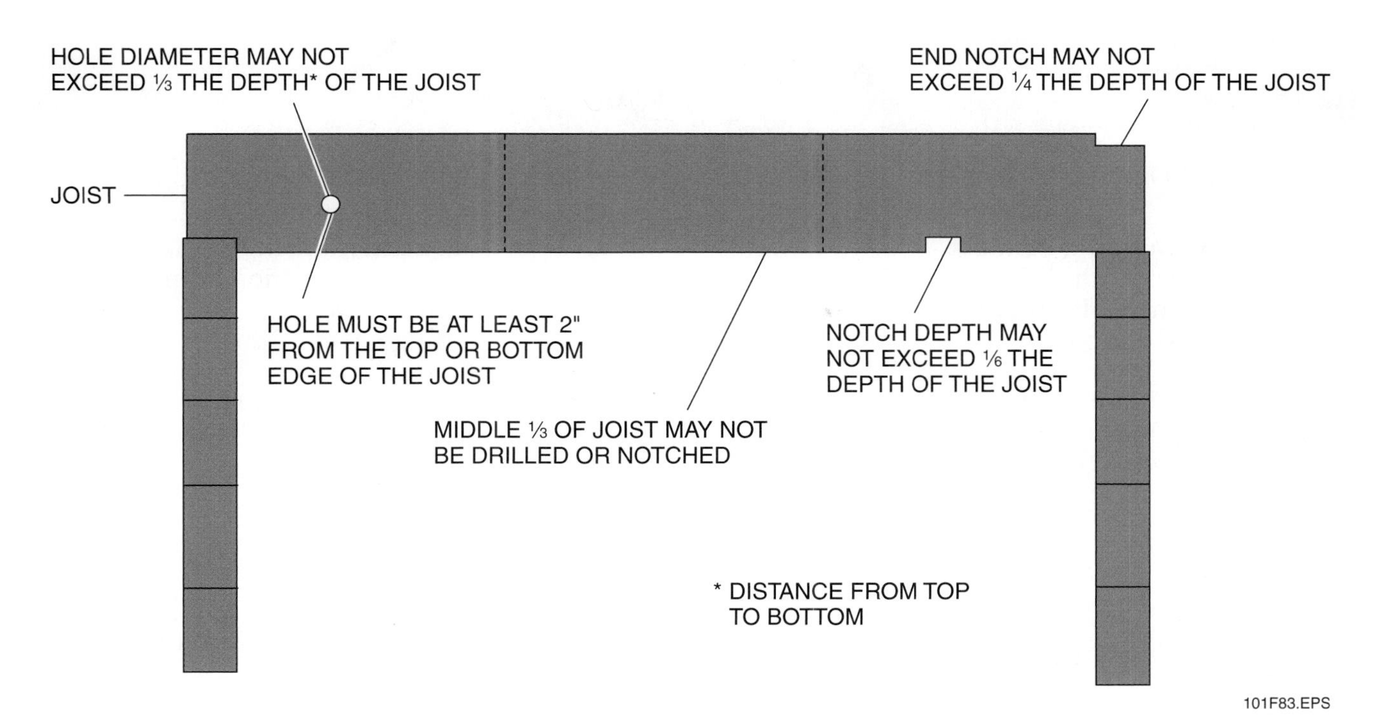

Figure 83 ◆ Be careful when notching structural members to preserve structural integrity.

One such area is under bathtubs where the weight of the tub may already be heavier than the joists are designed to hold.

Another common problem is compression or disturbance of insulation in walls by various trades. The energy performance of batt insulation is significantly reduced when it is compressed to fit in spaces without cutting or when gaps are left in wall cavities. It is also compromised if the insulation is wet or otherwise contaminated. Check all insulation before drywall is hung. This allows time for it to be fixed while the walls are still open. During the project, speak up if you see anything that might compromise the project's LEED goals. Make your supervisor aware if you see problems with stormwater management, dust containment, or any other protective measures. If you see others doing things incorrectly, talk with them to find out why. Encourage them to follow the LEED requirements to meet the project goals.

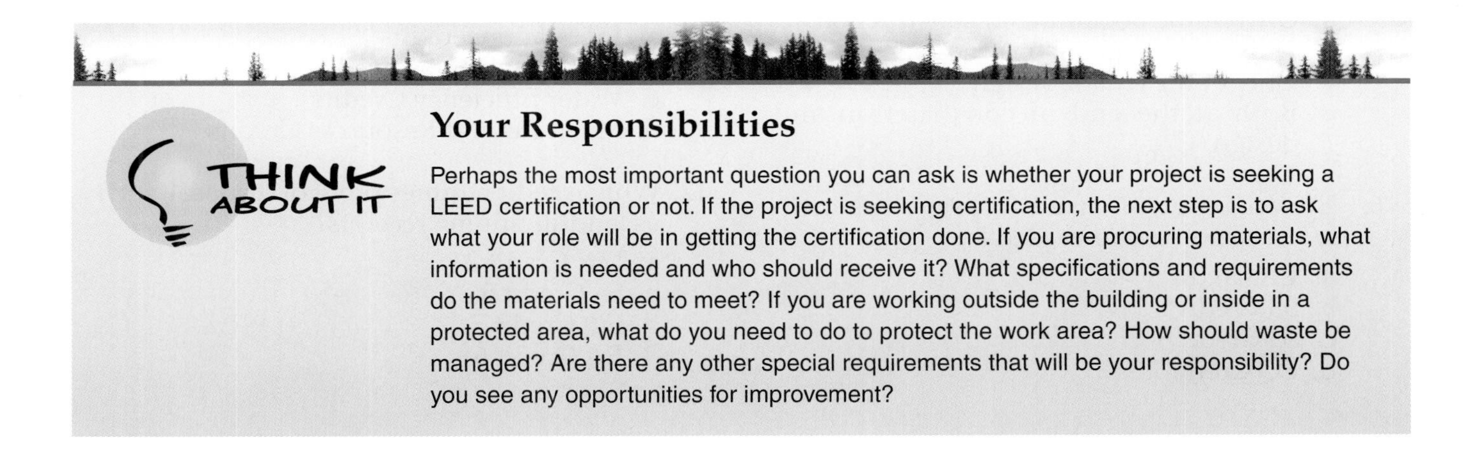

Your Responsibilities

Perhaps the most important question you can ask is whether your project is seeking a LEED certification or not. If the project is seeking certification, the next step is to ask what your role will be in getting the certification done. If you are procuring materials, what information is needed and who should receive it? What specifications and requirements do the materials need to meet? If you are working outside the building or inside in a protected area, what do you need to do to protect the work area? How should waste be managed? Are there any other special requirements that will be your responsibility? Do you see any opportunities for improvement?

Review Questions

1. The LEED rating system was developed by the _____.
   a. Lean Building Institute
   b. U.S. Green Building Council
   c. Sustainable Built Environment Initiative
   d. United Kingdom

2. The LEED rating system consists of a series of performance goals and requirements in _____ different categories.
   a. four
   b. five
   c. six
   d. seven

3. The LEED rating system that is for the individual tenant spaces in commercial buildings is _____.
   a. LEED-EB
   b. LEED-NC
   c. LEED-CI
   d. LEED-CS

4. How many points are there in the LEED-NC rating system?
   a. 21
   b. 45
   c. 69
   d. 110

5. When can a project team submit documentation for LEED credits?
   a. Only at the beginning of construction
   b. Only at the end of construction
   c. Only at the end of design
   d. Both at the end of construction and design

6. Erosion controls used to prevent stormwater runoff are rewarded by _____.
   a. EA Credit 5
   b. SS Credit 1
   c. SS Credit 6
   d. EQ Credit 2

7. A prerequisite for all LEED projects is to meet all the requirements of the _____.
   a. Lean Construction Institute
   b. EPA General Construction Permit
   c. LEED requirements
   d. local building code

8. During construction, trees should be protected with _____ around the trunk of the tree.
   a. tree protection fencing
   b. trenching
   c. erosion controls
   d. bark wrap

9. Construction projects can earn a Materials and Resources Credit by _____.
   a. using graywater
   b. using carbon monoxide monitors
   c. sending sewage to onsite treatment systems
   d. separating construction waste for recycling

10. Materials should *not* be installed in a building if they are _____.
    a. inefficient
    b. unsustainable
    c. damp or wet
    d. nonrenewable

11. The LEED credits that reward projects for finding new ways to be green are the _____.
    a. Innovation Credits
    b. Sustainable Sites Credits
    c. Water Efficiency Credits
    d. Materials and Resources Credits

12. Which credit requires proof of disposal, such as landfill tipping receipts?
    a. MR Credit 2
    b. SS Credit 5
    c. EQ Credit 5
    d. EA Credit 6

Review Questions

13. A common pitfall to LEED certification during construction is _____.
    a. lack of coordination with other trades
    b. smoking on the job site
    c. not following directions
    d. using nonrenewable materials

14. Thermal bridging can create areas where _____ can condense.
    a. air
    b. moisture
    c. solvents
    d. carbon

15. When installing insulation be sure to _____.
    a. only use fiberglass batting
    b. compress the batting to fit spaces
    c. cut the batting to fit spaces
    d. only use spray-in foam

Summary

As the green environment changes, the construction industry must also change. Past practices have had multiple negative impacts on the green environment. Today, there are many things that can be done to change for the better.

You can estimate your individual impact by calculating your carbon footprint. Your carbon footprint measures how much carbon your activities emit to the atmosphere. Some of the ways to reduce your carbon footprint include reducing product and energy use, recycling, planting vegetation, finding better sources of energy, and buying carbon offsets to help other people improve. As you consider your options, be sure to look for leverage points—small actions you can take to make a big difference.

As a craft worker, there are six major areas where you can improve a construction project's impact on the green environment. These include the site and landscape, water and wastewater, energy, materials and waste, indoor environment, and integrated strategies. Within each of these areas, you can have the most impact by eliminating the unnecessary uses of resources and finding more efficient ways to use resources. Look for better sources for resources, and better sinks for waste. Seek out integrated strategies where one action has multiple benefits. Again, look for leverage points, where small actions can make a big difference.

The LEED Green Building Rating System provides a useful measuring stick for industry improvement. LEED has seven categories of credits (Sustainable Sites, Water Efficiency, Energy and Atmosphere, Materials and Resources, Indoor Environmental Quality, Innovation in Design, and Regional Priority). Projects can be certified at four levels: Certified, Silver, Gold, and Platinum. Platinum is the most difficult rating level. LEED projects must meet certain basic requirements called prerequisites, and then can choose from points in the seven categories to achieve a rating level.

As a craft worker, you have many opportunities to contribute to a project's LEED certification. How you manage the project site is important. Pay attention to stormwater management and control systems. Be aware of how you protect building systems from potential indoor environmental contamination. Be careful to procure materials that meet specifications. Look for better sources and sinks for your materials and waste streams. Remember, your actions can make a real difference.

Notes

Notes

Notes

Trade Terms Introduced in This Module

Absorptive finish: A surface finish that will absorb dust, particles, fumes, and sound.

Acidification: A process that converts air pollution into acid substances. This leads to acid rain. Acid rain is best known for the damage it causes to forests and lakes. Also refers to the outflow of acidic water from metal and coal mines.

Aerated autoclaved concrete (AAC): A lightweight, precast building material that provides structural strength, insulation, and fire resistance. Typical products include blocks, wall panels, and floor panels.

Albedo: The extent to which an object reflects light from the sun. It is a ratio with values from 0 to 1. A value of 0 is dark (low albedo). A value of 1 is light (high albedo).

Alternative fuel: Any material or substance that can be used as a fuel other than conventional fossil fuels. They produce less pollution than fossil fuels. These include biodiesel and ethanol.

Aquifer: An underground layer of water-bearing rock or soil from which groundwater can be usefully extracted using a water well.

Aquifer depletion: A situation in which the withdrawal of water from its underground source is greater than the rate of natural recharge.

Best management practice (BMP): A way to accomplish something with the least amount of effort to achieve the best results. This is based on repeatable procedures that have proven themselves over time for large numbers of people.

Bio-based: A material made from substances derived from living matter. It typically refers to modern materials that have undergone more extensive processing. Linoleum, cork, and bamboo are examples.

Biodegradable: Organic material such as plant and animal matter and other substances originating from living organisms. These are capable of being broken down into innocuous products by the action of microorganisms.

Biodiversity: A measure of the variety among organisms present in different ecosystems.

Biofuel: A solid, liquid, or gas fuel consisting of or derived from recently dead biological material. The most common source is plants.

Biomimicry: The act of imitating nature to create new solutions modeled after natural systems.

Bioswale: An engineered depression designed to accept and channel stormwater runoff. A bioswale uses natural methods to filter stormwater, such as vegetation and soil.

Blackwater: Water or sewage that contains fecal matter or sources of pathogens.

Brownfield: Property that contains the presence or potential presence of a hazardous substance, pollutant, or contaminant. Cleaning up and reinvesting in these properties reduces development pressures on undeveloped, open land and improves and protects the environment.

Building-integrated photovoltaics (BIPVs): Photovoltaic systems built into other types of building materials. See *photovoltaic*.

Byproducts: Another product derived from a manufacturing process or a chemical reaction. It is not the primary product or service being produced.

Carbon cycle: The movement of carbon between the biosphere, atmosphere, oceans, and geosphere of the Earth. It is a biogeochemical cycle. In the cycle, there are sinks, or stores, of carbon. There are also processes by which the various sinks exchange carbon.

Carbon footprint: A measure of the impact that human activities have on the environment. It is determined by the amount of greenhouse gases produced. It is measured in units of pounds or kilograms of carbon dioxide.

Carpool: An arrangement in which several people travel together in one vehicle. The people take turns driving and share in the cost.

Charrette: An intense meeting of project participants that can quickly generate design solutions by integrating the abilities and interests of a diverse group of people.

Chlorofluorocarbons (CFCs): A set of chemical compounds that deplete ozone. They are widely used as solvents, coolants, and propellants in aerosols. These chemicals are the main cause of ozone depletion in the stratosphere.

Cob construction: An ancient building method using hand-formed lumps of earth mixed with sand and straw.

Commissioning: A review process conducted by a third party that involves a detailed design review, testing and balancing of systems, and system turnover.

Compact fluorescent lamp (CFL): A fluorescent light bulb that is designed to fit in a normal light fixture. They use less energy and last longer than incandescent light bulbs.

Conservation: Using natural resources wisely and at a slower rate than normal.

Deconstruction: Taking a building apart with the intent of salvaging reusable materials.

Deforestation: The removal of trees without sufficient replanting.

Desertification: The creation of deserts through degradation of productive land in dry climates by human activities.

Downcycling: Recycling one material into a material of lesser quality. An example is the recycling of plastics into lower-grade composites.

EarthCraft: A residential green building program of the Greater Atlanta Home Builders Association in partnership with Southface Energy Institute. This program, founded in 1999, serves as a blueprint for energy and resource efficient homes. It is supported by manufacturing and service industry partners.

Ecological footprint: A measure of the human demand on the Earth's ecosystems and natural environment. It compares human consumption of natural resources with the Earth's capacity to regenerate them. Measured in units of area such as hectares or acres.

Ecosystem: A combination of all plants, animals and microorganisms in an area that complement each other. These function together with all of the nonliving physical factors of the environment.

Embodied energy: The total energy required to bring a product to market, from raw material extraction to manufacturing to final transport and installation.

Endangered species: A population of a species at risk of becoming extinct. A threatened species is any species that is vulnerable to extinction in the near future.

Energy efficiency: Getting more use out of the electricity already generated.

Energy Star®: A United States government program to promote energy-efficient consumer products. It is a joint program of the U.S. Environmental Protection Agency and the U.S. Department of Energy.

Equipment idling: The operation of equipment while it is not in motion or performing work. Limiting idle times reduces air pollution and greenhouse gas emissions.

Erosion: The displacement of solids by wind, water, ice, or gravity or by living organisms. These solids include rocks and soil particles.

Flushout: Using fresh air in the building HVAC system to remove contaminants from the building.

FSC Certified: Wood or wood products that have met the Forest Stewardship Council's tracking process for sustainable harvest.

Fugitive emissions: Pollutants released to the air other than those from stacks or vents. They are often due to equipment leaks, evaporative processes, and wind disturbances.

Geothermal: Heat that comes from within the Earth.

Global climate change: Changes in weather patterns and temperatures on a planetary scale. This may lead to a rise in sea levels, melting of polar ice caps, increased droughts, and other weather effects.

Graywater: A nonindustrial wastewater generated from domestic processes. These include laundry and bathing. Graywater comprises 50 to 80 percent of residential wastewater.

Green Seal Certified: A certification of a product to indicate its environmental friendliness. Green Seal is a group that works with manufacturers, industry sectors, purchasing groups, and government at all levels to green the production and purchasing chain. Founded in 1989, Green Seal provides science-based environmental certification standards.

Halons: Ozone-depleting compounds consisting of bromine, fluorine, and carbon. They were commonly used as fire extinguishing agents, both in built-in systems and in handheld portable fire extinguishers. Halon production in the U.S. ended on December 31, 1993.

Hybrid vehicle: A vehicle that uses two or more power sources for propulsion.

Hydrochlorofluorocarbons (HCFCs): A group of manmade compounds containing hydrogen, chlorine, fluorine, and carbon. They are used for refrigeration, aerosol propellants, foam manufacture, and air conditioning. They are broken down in the lowest part of the atmosphere and pose a smaller risk to the ozone layer than other types of refrigerants.

Hydrologic cycle: The circulation and conservation of Earth's water supply. The process has five phases: condensation, infiltration, runoff, evaporation, and precipitation.

Indoor air quality (IAQ): The content of interior air that could affect health and comfort of building occupants.

Insulating concrete form (ICF): Rigid forms that hold concrete in place during curing and remain in place afterwards. The forms serve as thermal insulation for concrete walls.

Integrated design: A collaborative design methodology that emphasizes the input of knowledge from several areas in the development of a complete design.

Just-in-time delivery: A material delivery strategy that reduces material inventory. The material that is delivered is used immediately.

Leadership in Energy and Environmental Design (LEED)®: A point-based rating system used to evaluate the environmental performance of buildings. Developed by the U.S. Green Building Council, LEED provides a suite of standards for environmentally sustainable construction. Founded in 1998, LEED focuses on the certification of commercial and residential buildings.

Life cycle: The useful life of a system, product, or building.

Life cycle analysis: An analytic technique to evaluate the environmental impact of a system, product, or building throughout its life cycle. This includes the extraction or harvesting of raw materials through processing, manufacture, installation, use, and ultimate disposal or recycling.

Life cycle cost: The cost of a system or a component over its entire life span.

Lithium ion (Li-ion): A type of rechargeable battery in which a lithium ion moves between the anode and cathode. They are commonly used in consumer electronics.

Local materials: Materials that come from within a certain number of miles from the project. Materials produced locally use less energy during transportation to the site. The LEED system of building certification offers points for the use of regional or local materials.

Low-emission vehicle: Vehicles that produce fewer emissions than the average vehicle. Beginning in 2001, all light vehicles sold nationally were required to meet this standard.

Minimum efficiency reporting value (MERV): A measurement scale designed in 1987 by the American Society of Heating, Refrigeration, and Air-Conditioning Engineers (ASHRAE) to rate the effectiveness of air filters. The scale is designed to represent the worst-case performance of a filter when dealing with particles in the range of 0.3 to 10 microns.

Monofill: A landfill that accepts only one type of waste, typically in bales.

Multi-function material: A material that can be used to perform more than one function in a facility.

Nano-material: A material with features smaller than a micron in at least one dimension.

Nickel cadmium (NiCad): A popular type of rechargeable battery using nickel oxide hydroxide and metallic cadmium as electrodes.

Nonrenewable: A material or energy source that cannot be replenished within a reasonable period.

Nontoxic: Substances that are not poisonous.

Offgassing: The evaporation of volatile chemicals at normal atmospheric pressure. Building materials release chemicals into the air through evaporation.

Overpopulation: Excessive population of an area to the point of overcrowding, depletion of natural resources, or environmental deterioration.

Ozone depletion: A slow, steady decline in the total amount of ozone in Earth's stratosphere.

Ozone hole: A large, seasonal decrease in stratospheric ozone over Earth's polar regions. The ozone hole does not go all the way through the layer.

Papercrete: A fiber-cement material that uses waste paper for fiber.

Passive survivability: The ability of a building to continue to offer basic function and habitability after a loss of infrastructure (i.e., water and power).

Pathogen: An infectious agent that causes disease or illness.

Payback period: The amount of time it takes to break even on an investment.

Peak shaving: An energy management strategy that reduces demand during peak times of the day and shifts it to off-peak times, such as at night.

Pervious concrete: A mixture of coarse aggregate, Portland cement, water, and little to no sand. It has a 15 to 25 percent void structure and allows 3 to 8 gallons of water per minute to pass through each square foot. Also known as permeable concrete.

Photovoltaic: A technology that converts light directly into electricity.

Pollution prevention: The prevention or reduction of pollution at the source.

Post-consumer: Products made out of material that has been used by the end consumer and then collected for recycling.

Post-industrial/pre-consumer: Material diverted from the waste stream during the manufacturing process.

Preservation: The act and advocacy of the protection of the natural environment.

Radio-frequency identification (RFID): An automatic identification method. It relies on storing and remotely retrieving data using devices called RFID tags.

Rainwater harvesting: The gathering and storing of rainwater. Systems can range from a simple barrel at the bottom of a downspout to multiple tanks with pumps and controls.

Rammed earth: An ancient building technique similar to adobe using soil that is mostly clay and sand. The difference is that the material is compressed or tamped into place, usually with forms that create very flat vertical surfaces.

Rapidly renewable: A material that is replenished by natural processes at a rate comparable to its rate of consumption. The LEED system of building certification offers points for rapidly renewable materials that regenerate in 10 years or less, such as bamboo, cork, wool, and straw.

Raw material: A material that has been harvested or extracted directly from nature. It is in an unprocessed or minimally processed state.

Recyclable: Material that still has useful physical or chemical properties after serving its original purpose. It can be reused or remanufactured into additional products. Plastic, paper, glass, used oil, and aluminum cans are examples of recyclable materials.

Recycled content: A material that contains components that would otherwise have been discarded.

Recycled plastic lumber (RPL): A wood-like product made from recovered plastic, either by itself or mixed with other materials. It can be used as a substitute for concrete, wood, and metals.

Recycling: The reprocessing of old materials into new products. A goal is to prevent the waste of potentially useful materials and reduce the consumption of fresh raw materials.

Renewable: A resource that may be naturally replenished.

Reusable: A material that can be used again without reprocessing. This can be for its original purpose or for a new purpose.

Salvaged: A discarded or damaged material that is saved from destruction or waste and put to further use.

Sedimentation: The process of depositing a solid material from a state of suspension in a fluid, usually air or water.

Sick Building Syndrome: A variety of illnesses thought to be caused by poor indoor air. Symptoms include headaches, fatigue, and other problems that increase with continued exposure.

Smart material: Materials that have one or more properties that can be significantly changed. These changes are driven by external stimuli.

Solar: Energy from the sun in the form of heat and light.

Solid waste: Products and materials discarded after use in homes, businesses, restaurants, schools, industrial plants, or elsewhere.

Solvent-based: A material that consists of particles suspended or dissolved in a solvent. A solvent is any substance that will dissolve another. Solvent-based building materials typically use chemicals other than water as their solvent, including toluene and turpentine, with hazardous health effects.

Sprawl: Unplanned development of open land.

Stormwater runoff: The unfiltered water that reaches streams, lakes, and oceans after a rainstorm by means of flowing across impervious surfaces.

Strawbale construction: A building method that uses straw as the structural element, insulation, or both. It has advantages over some conventional building systems because of its cost and availability.

Structural insulated panel (SIP): A composite building material used for exterior building envelopes. It consists of a sandwich of two layers of structural board with an insulating layer of foam in between. The board is usually oriented strand board (OSB) and the foam can be polystyrene, soy-based foam, urethane, or even compressed straw.

Sustainably harvested: A method of harvesting a material from a natural ecosystem without damaging the ability of the ecosystem to continue to produce the material indefinitely.

Takeback: A requirement that waste from packaging or products be recovered by the manufacturer or provider at the end of its life cycle.

Thermal bridging: A condition created when a thermally conductive material bypasses an insulation system, allowing the rapid flow of heat from one side of a building wall to the other. Metal components, including metal studs, nails, and window frames, are common culprits.

Thermal mass: A property of a material related to density that allows it to absorb heat from a heat source, and then release it slowly. Common materials used to provide thermal mass include adobe, mud, stones, or even tanks of water.

Urbanization: The removal of the rural characteristics of an area. A redistribution of populations from rural to urban settlements.

Urea formaldehyde: A transparent thermosetting resin or plastic. It is made from urea and formaldehyde heated in the presence of a mild base. Urea formaldehyde has negative effects on human health when allowed to offgas or burn.

Vapor-resistant: A material that resists the flow of water vapor.

Virgin material: A material that has not been previously used or consumed. It also has not been subjected to processing. See also *raw material.*

Volatile organic compounds (VOCs): Gases that are emitted over time from certain solids or liquids. Concentrations of many VOCs are up to 10 times higher indoors than outdoors. Examples include paints and lacquers, paint strippers, cleaning supplies, pesticides, building materials, and furnishings.

Walk-off mat: Mats in entry areas that capture dirt and other particles.

Waste separation: Separating waste into recyclable and nonrecyclable materials.

Water efficiency: The planned management of potable water to prevent waste, overuse, and exploitation of the resource. Includes using less water to achieve the same benefits.

Water-resistant: A material that hinders the penetration of water.

Water-based: Materials that use water as a solvent or vehicle of application.

Waterproof: A material that is impervious to or unaffected by water.

Wetland: Lands where saturation with water is the dominant factor. This determines the way soil develops and the types of plant and animal communities living in the soil and on its surface.

Xeriscaping: Landscaping that requires little or no water for irrigation. Xeriscaping can be achieved through smart plant selection, mulching, and other tactics.

Common Acronyms

AAC aerated autoclaved concrete
ACH air changes per hour
BIPV building-integrated photovoltaic
BMP best management practice
BTU British thermal unit
CFC chlorofluorocarbon
CFL compact fluorescent lamp
CO carbon monoxide
CO_2 carbon dioxide
EA Energy and Atmosphere (LEED credit category)
EEM energy efficient mortgage
EMF electromagnetic field
EPA Environmental Protection Agency
EQ Indoor Environmental Quality (LEED credit category)
ERV energy recovery ventilator
FSC Forest Stewardship Council
GHG greenhouse gas
HCFC hydrochlorofluorocarbon
HERS home energy rating system
HID high intensity discharge
HVFA high volume flyash
IAQ indoor air quality
ICF insulating concrete form
ID Innovation in Design (LEED credit category)
IDP integrated design process
LCA life cycle assessment
LEED® Leadership in Energy and Environmental Design®
LEED-AP Leadership in Energy and Environmental Design Accredited Professional
LEED-CI Leadership in Energy and Environmental Design for Commercial Interiors
LEED-CS Leadership in Energy and Environmental Design for Core and Shell
LEED-EB Leadership in Energy and Environmental Design for Existing Buildings
LEED-H Leadership in Energy and Environmental Design for Homes
LEED-NC Leadership in Energy and Environmental Design for New Construction and Major Renovation
LEED-ND Leadership in Energy and Environmental Design for Neighborhood Development
Li-ion lithium ion
Low-E low-emissivity
MCS multiple chemical sensitivity
MDF medium-density fiberboard
MERV minimum efficiency reporting value
MR Materials and Resources (LEED credit category)
N_2O nitrous oxide
NACH natural air changes per hour
NAHB National Association of Home Builders
NFRC National Fenestration Rating Council
NiCad nickel cadmium
OPC off-peak cooling
OSB oriented strand board
OVE optimum value engineering
PB particleboard
PF phenyl formaldehyde
PV photovoltaic
PVC polyvinyl chloride
REC Renewable Energy Credit
RFID radio-frequency identification
RP Regional Priority (LEED credit category)
RPL recycled plastic lumber
R-value resistance to heat flow
SFI Sustainable Forest Initiative
SHGC solar heat gain coefficient
SIP structural insulated panel
SS Sustainable Sites (LEED credit category)
UF urea formaldehyde
USGBC United States Green Building Council
UV ultraviolet
U-value resistance to heat loss (also known as U-factor)
VOC volatile organic compound
WE Water Efficiency (LEED credit category)

Appendix A

LEED for New Construction and Major Renovation, Version 3 Registered Project Checklist

SUSTAINABLE SITES		26 POINTS
Prereq 1	**Construction Activity Pollution Prevention**	Required
Credit 1	**Site Selection**	1
Credit 2	**Development Density and Community Connectivity**	5
Credit 3	**Brownfield Redevelopment**	1
Credit 4.1	**Alternative Transportation**, Public Transportation Access	6
Credit 4.2	**Alternative Transportation**, Bicycle Storage and Changing Rooms	1
Credit 4.3	**Alternative Transportation**, Low-Emitting and Fuel-Efficient Vehicles	3
Credit 4.4	**Alternative Transportation**, Parking Capacity	2
Credit 5.1	**Site Development**, Protect or Restore Habitat	1
Credit 5.2	**Site Development**, Maximize Open Space	1
Credit 6.1	**Stormwater Design**, Quantity Control	1
Credit 6.2	**Stormwater Design**, Quality Control	1
Credit 7.1	**Heat Island Effect**, Nonroof	1
Credit 7.2	**Heat Island Effect**, Roof	1
Credit 8	**Light Pollution Reduction**	1

WATER EFFICIENCY		10 Points
Prereq 1	**Water Use Reduction**	Required
Credit 1	**Water Efficient Landscaping**	2 to 4
	Reduce by 50%	2
	No Potable Use or Irrigation	4
Credit 2	**Innovative Wastewater Technologies**	2
Credit 3	**Water Use Reduction**	2 to 4
	30% Reduction	2
	35% Reduction	3
	40% Reduction	4

ENERGY & ATMOSPHERE		35 POINTS
Prereq 1	**Fundamental Commissioning of the Building Energy Systems**	Required
Prereq 2	**Minimum Energy Performance**	Required
Prereq 3	**Fundamental Refrigerant Management**	Required
Credit 1	**Optimize Energy Performance** (Number of Credits Based on %)	1 to 19
Credit 2	**On-Site Renewable Energy** (Number of Credits Based on %)	1 to 7
Credit 3	**Enhanced Commissioning**	2
Credit 4	**Enhanced Refrigerant Management**	2
Credit 5	**Measurement & Verification**	3
Credit 6	**Green Power**	2

MATERIALS & RESOURCES		14 POINTS
Prereq 1	**Storage & Collection of Recyclables**	Required
Credit 1.1	**Building Reuse**	1 to 3
	Maintain 55% of Existing Walls, Floors, and Roof	1
	Maintain 75% of Existing Walls, Floors, and Roof	2
	Maintain 95% of Existing Walls, Floors, and Roof	3
Credit 1.2	**Building Reuse**, Maintain Interior Nonstructural Elements	1
Credit 2	**Construction Waste Management**	1 to 2
	50% Recycled or Salvaged	1
	75% Recycled or Salvaged	2
Credit 3	**Materials Reuse**	1 to 2
	Reuse 5%	1
	Reuse 10%	2

MATERIALS & RESOURCES (Continued)		**14 POINTS**
Credit 4	**Recycled Content**	1 to 2
	10% of Content	1
	20% of Content	2
Credit 5	**Regional Materials**	1 to 2
	10% of Materials	1
	20% of Materials	2
Credit 6	**Rapidly Renewable Materials**	1
Credit 7	**Certified Wood**	1

INDOOR ENVIRONMENTAL QUALITY		**15 POINTS**
Prereq 1	**Minimum Indoor Air Quality Performance**	Required
Prereq 2	**Environmental Tobacco Smoke (ETS) Control**	Required
Credit 1	**Outdoor Air Delivery Monitoring**	1
Credit 2	**Increased Ventilation**	1
Credit 3.1	**Construction Indoor Air Quality Management Plan**, During Construction	1
Credit 3.2	**Construction Indoor Air Quality Management Plan**, Before Occupancy	1
Credit 4.1	**Low-Emitting Materials**, Adhesives and Sealants	1
Credit 4.2	**Low-Emitting Materials**, Paints and Coatings	1
Credit 4.3	**Low-Emitting Materials**, Flooring Systems	1
Credit 4.4	**Low-Emitting Materials**, Composite Wood and Agrifiber Products	1
Credit 5	**Indoor Chemical and Pollutant Source Control**	1
Credit 6.1	**Controllability of Systems**, Lighting	1
Credit 6.2	**Controllability of Systems**, Thermal Comfort	1
Credit 7.1	**Thermal Comfort**, Design	1
Credit 7.2	**Thermal Comfort**, Verification	1
Credit 8.1	**Daylight and Views**, Daylight	1
Credit 8.2	**Daylight and Views**, Views	1

INNOVATION IN DESIGN		**6 POINTS**
Credit 1	**Innovation in Design**	1 to 5
	Innovation or Exemplary Performance	1
	Innovation or Exemplary Performance	1
	Innovation or Exemplary Performance	1
	Innovation	1
	Innovation	1
Credit 2	**LEED® Accredited Professional**	1

REGIONAL PRIORITY		**4 POINTS**
Credit 1	**Regional Priority**	1 to 4
	Regionally Defined Credit Achieved	1
	Regionally Defined Credit Achieved	1
	Regionally Defined Credit Achieved	1
	Regionally Defined Credit Achieved	1

PROJECT TOTALS (Certification Estimates) **110 POINTS**

Certified: 40–49 points, **Silver:** 50–59 points, **Gold:** 60–79 points, **Platinum:** 80 points and above

Appendix B

Worksheet 1: Inventory Your Household Impacts

WORKSHEET 1: INVENTORY YOUR HOUSEHOLD IMPACTS

Conducting an inventory of your household consumption, waste generation, and activities is the first step in understanding how you can reduce your impact. Answer the following questions based on your best guess. If you share a household with other people, divide the total answer for your household by the number of people who live in your home.

How many gallons of garbage do you put out each week? An average garbage can holds 32 gallons. Multiply by 52 to calculate the gallons of garbage you throw away per year.

Gallons of garbage per year: ____________________________

How much electricity do you use per year? You'll find this information on your electricity bill. You can add up the total for 12 months worth of bills, or multiply a monthly average by 12. If you know how much your electricity costs per month, you can estimate the amount of energy used in kilowatt-hours by dividing your total bill by 10.

Total electricity per year in kilowatt-hours: ____________________________

How many therms of natural gas do you use per year? Check your natural gas bill if you get one. It will tell you how many therms of gas you use per month. Your highest values will probably be during the winter heating season. If you know your monthly average, multiply by 12 to calculate your annual use.

Total therms of natural gas per year: ____________________________

How many gallons of propane do you use each year? You can check this on your propane bill if you get one. Add up the total number of gallons per year.

Total gallons of propane per year: ____________________________

How many gallons of fuel oil do you use per year? You can check this on your fuel oil bill if you get one. Add up the total number of gallons per year.

Total gallons of fuel oil per year: ____________________________

On average, what is your monthly combined water and sewage bill? Check your monthly bill if you get one and add up all the amounts for a one-year period. If you don't get a water and sewage bill, you can estimate this amount as approximately $75 per person per year.

Annual cost for water and sewage: ____________________________

How many square feet is your house or dwelling? If you don't know, draw a floor plan of your house or apartment and estimate the floor area in square feet.

Size of house (floor area) in square feet: ____________________________

101WS01A.EPS

WORKSHEET 1 (Continued)

On average, how many miles do you drive your household vehicles per week, and what are their average fuel efficiencies? 300 miles per week per vehicle or 15,000 miles per year is about average in the United States. If you're not sure about fuel efficiency, assume your car gets 22 miles to the gallon, which is about average. If you know how many gallons of fuel you use per week, skip directly to the end of the line.

Car 1 miles per week ___________ / *Car 1 miles per gallon* ___________ = *Car 1 gallons per week* ___________

Car 2 miles per week ___________ / *Car 2 miles per gallon* ___________ = *Car 2 gallons per week* ___________

Car 3 miles per week ___________ / *Car 3 miles per gallon* ___________ = *Car 3 gallons per week* ___________

Add up the gallons of gas per week for all your vehicles, and then multiply by 52 to estimate the gallons of gasoline you use per year.

Total gallons of gasoline you use per year: ___________________________

On average, how much do you travel each year on airplanes? Estimate the number of flight segments below for each of the three distances. A round trip counts as two segments, and each segment of a multi-segment flight counts as its own flight. For example, if you fly from Baltimore to Cincinnati to Atlanta, that counts as two flight segments.

Number of short-haul flight segments (less than 700 miles or 2 hours): ______

Number of medium-haul flight segments (700 – 2,500 miles or 2 – 4 hours): ______

Number of long-haul flight segments (more than 2,500 miles or longer than 4 hours): ______

On average, how many miles do you travel on public transportation per year?

Number of miles per year on transit bus/subway: ______

Number of miles per year on intercity bus: ______

Number of miles per year on intercity train: ______

Add up the total number of miles per year on public transportation: ______________________

Your answers to these questions will help you calculate your own carbon footprint later in the module. Keep track of your answers in the workbook or on a separate worksheet.

101WS01B.EPS

Appendix C

Worksheet 2: Inventory Your Product Impacts

WORKSHEET 2: INVENTORY YOUR PRODUCT IMPACTS

Consider the products you buy or that are bought for you over the course of a year. Answer the following questions based on your best guess. If you share a household with other people, divide the total answer for your household by the number of people who live in your home.

Eating out:	$ ________ *per month* × 12 = $ ________ *per year*
Meat, fish, & protein:	$ ________ *per month* × 12 = $ ________ *per year*
Cereals & baked goods:	$ ________ *per month* × 12 = $ ________ *per year*
Dairy:	$ ________ *per month* × 12 = $ ________ *per year*
Fruits & vegetables:	$ ________ *per month* × 12 = $ ________ *per year*
Other:	$ ________ *per month* × 12 = $ ________ *per year*

On average each month, how much do you spend on the following goods and services? Multiply each amount by 12 to estimate your average annual spending in each category. If you already know how much you spend per year, you can skip the monthly amount and write the average amount per year at the end of each line.

Clothing:	$ ________ *per month* × 12 = $ ________ *per year*
Furnishings & household items:	$ ________ *per month* × 12 = $ ________ *per year*
Other goods:	$ ________ *per month* × 12 = $ ________ *per year*
Services:	$ ________ *per month* × 12 = $ ________ *per year*

101WS02.EPS

Worksheet 3: Determine your Carbon Footprint

WORKSHEET 3: DETERMINE YOUR CARBON FOOTPRINT

Category	Amount	Carbon Factor	Total Pounds of CO_2/Year
Gallons of garbage/year:	________	*× 2 lbs/gallon =*	________*lbs/year*
Total electricity/year in kilowatt-hours:	________	*× 1.4 lbs/kWh =*	________*lbs/year*
Total therms of natural gas/year:	________	*× 11.7 lbs/gallon =*	________*lbs/year*
Total gallons of propane/year:	________	*× 12.7 lbs/gallon =*	________*lbs/year*
Total gallons of fuel oil/year:	________	*× 22.4 lbs/gallon =*	________*lbs/year*
Annual cost for water and sewage:	________	*× 8.9 lbs/$ =*	________*lbs/year*
House size (floor area) in square feet:	________	*× 2.1 lbs/sq ft =*	________*lbs/year*
Gallons of gasoline/year:	________	*× 20 lbs/gallon =*	________*lbs/year*
No. of short-haul flight segments/year:	________	*× 304 lbs/segment (avg) =*	________*lbs/year*
No. of medium-haul flight segments/year:	________	*× 726 lbs/segment (avg) =*	________*lbs/year*
No. of long-haul flight segments/year:	________	*× 2,217 lbs/segment (avg) =*	________*lbs/year*
Miles per year on public transportation:	________	*× 0.5 lb/mile =*	________*lbs/year*
Eating out (dollars/year):	________	*× 0.8 lb/dollar =*	________*lbs/year*
Meat, fish, & protein (dollars/year):	________	*× 3.2 lbs/dollar =*	________*lbs/year*
Cereals & baked goods (dollars/year):	________	*× 1.6 lbs/dollar =*	________*lbs/year*
Dairy (dollars/year):	________	*× 4.2 lbs/dollar =*	________*lbs/year*
Fruits & vegetables (dollars/year):	________	*× 2.6 lbs/dollar =*	________*lbs/year*
Other (dollars/year):	________	*× 1.0 lb/dollar =*	________*lbs/year*
Clothing (dollars/year):	________	*× 1.0 lb/dollar =*	________*lbs/year*
Household items (dollars/year):	________	*× 1.0 lb/dollar =*	________*lbs/year*
Other goods (dollars/year):	________	*× 0.75 lb/dollar =*	________*lbs/year*
Services (dollars/year):	________	*× 0.4 lb/dollar =*	________*lbs/year*
Total =			________________**lbs/year**

101WS03.EPS

Additional Resources

Additional Resources

This module is intended to be a thorough resource for task training. The following reference works are suggested for further study. These are optional materials for continued education rather than for task training.

Field Guide for Sustainable Construction. Department of Defense – Pentagon Renovation and Construction Office. (2004). PDF, 2.6 MB, 312 pgs. Available for download at http://renovation.pentagon.mil.

Greening Federal Facilities, 2nd ed. U.S. Department of Energy Federal Energy Management Program. (2001). PDF, 2.1 MB, 211 pgs. Available for download at www.ofee.gov.

Natural Capitalism. Lovins, A., Hawkin, P., and Lovins, L.H. (1995). Little, Brown, & Company, Boston, MA. Available online at www.natcap.org.

Sustainable Buildings Technical Manual. Public Technologies, Inc./U.S. Department of Energy. (2006). PDF, 3.1 MB, 292 pgs. Available for download at www.smartcommunities.ncat.org.

Sustainable Construction: Green Building Design and Delivery, 2nd ed. Kibert, C.J. (2007). Wiley, New York, NY.

The HOK Guidebook to Sustainable Design, 2nd ed. Mendler, S.F., Odell, W., and Lazarus, M.A. (2005). Wiley, New York, NY.

References

The following list is a compilation of websites referenced in this module.

Arid Solutions Inc.: www.aridsolutionsinc.com

Carbon Footprint: www.carbonfootprint.com

Database of State Incentives for Renewables & Efficiency: www.dsireusa.org

Energy Star: www.energystar.gov

Forest Stewardship Council: www.fsc.org

Green Building Certification Institute: www.gbci.org

Green Globes: www.greenglobes.com

Green Seal: www.greenseal.org

Green-e: www.green-e.org

Habitat for Humanity: www.habitat.org

International Initiative for a Sustainable Built Environment: www.iisbe.org

NAHB Research Center: www.nahbrc.org

Natural Capitalism: www.natcap.org

Office of the Federal Environmental Executive: www.ofee.gov

Refining Process: www.myfootprint.org

Smart Communities Network: www.smartcommunities.ncat.org

The Carpet and Rug Institute: www.carpet-rug.org

The PLANTS Database: www.plants.usda.gov

U.S. Environmental Protection Agency: Clean Energy www.epa.gov/cleanenergy

U.S. Green Building Council: www.leedbuilding.org

U.S. Green Building Council: www.usgbc.org

WaterSense: www.epa.gov/watersense

Acknowledgments

Figure Credits

Photos.com, © 2008 Jupiterimages Corporation, Cover, page iii, page vii, 101F01–101F05, 101SA01, 101SA02, 101SA04, 101SA06, 101SA08, 101SA09, 101SA10, 101SA12, 101SA14, 101SA15, 101F49B, 101F50

Virginia Tech, 101F06, 101SA03, 101F09, 101SA05, 101F11–101F13, 101F16, 101F17–101F19, 101F21, 101F23, 101F24, 101F25B, 101F26, 101F27, 101F29, 101F30, 101SA11, 101F31, 101F32, 101F35, 101SA13, 101F36–101F46, 101F48, 101F49A, 101F51–101F53, 101F56–101F65, 101F68–101F78, 101SA16, 101F80, 101SA17, 101SA18, 101F81

National Oceanic and Atmospheric Administration, 101F07

The City of Santa Monica, 101F08

Energy Information Administration, U.S. Department of Energy, 101F10

Office of Energy Efficiency and Renewable Energy, U.S. Department of Energy, 101F15

Waterless Co., 101F20

Brac Systems, 101F22

Tim Davis, 101SA07, 101F25A, 101SA16

Courtesy of the Center for Resource Solutions, www.Green-E.org, 101F28

Courtesy USDA-NRCS, 101F33

Images provided by Tate Access Floors Inc., 101F47, 101F67

© U.S. Green Building Council, 101F54

Megan Paye, 101F55

Arid Solutions Inc., www.aridsolutionsinc.com, 101F66

Green Advantage, 101F79

Ren Solutions, 101F82

Tim Dean, 101SA19

Green Advantage® Study Guide

Appendix E

Your Role in the Green Environment

70101-08

GREEN ADVANTAGE® STUDY GUIDE

This study guide compliments the material you learned in this module. It presents additional vocabulary words that are included in the Green Advantage certification exam, along with sample questions. The questions are formatted similar to how they will appear on the certification test. Familiarize yourself with the content as well as the format of the questions so you can be best prepared for the test.

Each section of this guide highlights concepts that are important in green building and are covered in the Green Advantage certification exam. When using the guide to prepare for the exam, it is suggested that you read each section and note the important concepts and vocabulary.

General Test Taking Tips

When taking the Green Advantage exam, read each question carefully and note the wording that is used. Common phrases throughout the test use qualifiers and other distracters including:

Except which…

Least effective…

Saves "water" vs. "energy"

Green Building Standards (relevant organizations)

Energy Star® – energy efficiency

Water uses – little to no (i.e. most likely to save water)

INTRODUCTION TO GREEN BUILDING

Benefits of Green Building

Buildings consume a large portion of raw materials, energy, and water. The nature of a green building is to provide an alternative to the use of such large quantities of raw materials and energy. Builders of green buildings look to use materials more efficiently by incorporating recycled materials into the construction as well as using alternative materials. The design of green buildings helps to reduce the energy consumption of the facility to include less electricity use and less water. Based upon the wise use of materials and careful design, these building produce less pollution than typical buildings.

The reduction in pollution is associated with both indoor and outdoor air quality. By using less energy to operate the building, harmful greenhouse gas emissions are reduced. Indoor air quality is important for occupant comfort, health, and performance. By using materials that produce less volatile compounds and give off lower levels of pollution, the indoor air is improved.

Life Cycle Cost Benefits

When viewing a building from a life cycle perspective, all aspects associated with the construction, operation, and maintenance of a facility are investigated. Each aspect of the building is assessed according to the damage that it may have on the environment over the life of the building. This assessment includes reviewing the raw materials from the point of extraction used in production of the material and the energy associated with their production. Looking at the life cycle does not stop there. It includes the distribution of the material, the use of the material, and disposal or recycling of the material, as well as all of the transportation costs that are a part of moving the material. The life cycle considers all the energy used from the time the material is produced to the time it is taken out of service. (This includes how the building occupants are affected and if any considerations to other material or building systems need to be made.)

Impact and Issues Regarding the Environment

Each of the following concepts relate to issues that impact the environment. They include theories and development strategies that are aimed at increasing the awareness of the impact building has on the environment. Also highlighted are certifying entities of green building practices and resources for contractors who must comply with new standards. More specific information about the services these entities provide is listed on their respective websites. The links are provided in the guide and you are encouraged to review these in preparation for the exam. Additionally, energy and green building legislation is discussed. This type of legislation will change over time. When you prepare for your exam, be certain to investigate the websites listed in this guide to be sure you are aware of the most current legislation.

Peak Oil

Peak oil is most commonly defined as the point in time when the maximum rate of global petroleum extraction is reached. Once this point is reached, the rate of petroleum production enters a state of permanent decline. The term *peak oil* does not refer to a time where oil runs out. Peak oil addresses the rate of petroleum production. There are various predictions for when this will occur. According to Green Advantage, this will occur in the next 5 to 20 years.

Increased Surface Temperatures

Global average surface temperatures have risen approximately 1°F in the last 100 years. If atmospheric CO_2 concentrations are allowed to increase to more than double or triple the current levels, the maximum projected increase in global average surface temperatures may increase by 9–10 degrees by the year 2100. Additional information regarding the monitoring of the increase in surface temperatures around the globe can be found at the Goddard Institute for Space Studies website (*http://data.giss.nasa.gov*).

Transit-oriented development is the creation of compact, walkable communities centered around high quality transit systems. This makes it possible for residents to not be completely dependent on a car to get to work, shopping, or recreation activities. This type of development can help reduce global warming as less fossil fuel will be combusted by the public. This type of development is the opposite of urban sprawl and reduces the impact of development. It will help to reduce the health hazards created by urban sprawl, such as obesity and outdoor air pollution.

Resources for Constructors and Certifying Entities

There are several organizations designed to help builders make decisions about building more sustainably. Some of these organizations and their associated missions are highlighted below. *All of the information provided about these organizations is taken directly from their websites and the links to their websites are included.* It is suggested that in your preparation for the Green Advantage exam that you refer to these websites.

The following is a brief listing of resources for constructors that are mentioned by Green Advantage.

- **BuildingGreen, LLC** is an independent company committed to providing accurate, unbiased, and timely information designed to help building-industry professionals and policy makers improve the environmental performance and reduce the adverse impacts of buildings.

  The corporate mission of BuildingGreen, LLC is to facilitate transformation of the North American building industry into a force for local, regional, and global environmental protection; for preservation and restoration of the natural environment; and for creation of healthy indoor environments while promoting the well-being of the company and its employees, owners, and associates. More information about the company can be found on their website at *www.buildinggreen.com*.
- The **Whole Building Design Group (WDBG)** is the only web-based portal providing government and industry practitioners with one-stop access to up-to-date information on a wide range of building-related guidance, criteria, and technology from a whole buildings perspective. The organization is divided into three major categories—Design Guidance, Project Management, and Operations & Maintenance. The WBDG website contains Resource Pages that offer reductive summaries on particular topics.

  The WBDG website is offered as an assistance to the building community by the National Institute of Building Sciences (NIBS) through funding support from the Department of Defense, the NAVFAC Engineering Innovation and Criteria Office, the Army Corps of Engineers, the U.S. Air Force, the U.S. General Services Administration (GSA), the Department of Veterans Affairs, the National Aeronautics and Space Administration (NASA), and the Department of Energy, with the assistance of the Sustainable Buildings Industry Council (SBIC). A Board of Direction and Advisory Committee, consisting of representatives from over 25 participating federal agencies, guides the development of the WBDG. More information can be found on their website at *www.wbdg.org*.
- **Greenerbuilding** is a non-profit website whose sole mission is to encourage the construction of better buildings. Throughout the site, you'll find building products, local dealers, and helpful programs near you. Greenerbuilding.org is a place for individuals to share their personal expertise about products with the sustainable building community. More information can be found on their website at *www.greenerbuilding.org*.

The following are current national certification programs highlighted in the Green Advantage material.

- The American Lung Association **Health House®** Program is an audit program for builders and consumers to test and verify that a house has met the climate-specific Health House® performance standards. The Health House® building program is currently the only residential building program that combines health, energy, and resource efficiency into one product. The Health House® components include a builder training and marketing program and a homeowner education program. More information can be found on their website at *www.healthhouse.org*.
- The U.S. Department of Energy (DOE) **Building America** program advocates a systems engineering approach to home building that unites segments of the building industry which traditionally work independently of one another. It forms teams of architects, engineers, builders, equipment manufacturers, material suppliers, community planners, mortgage lenders, and contractor trades. More information can be found on their website at *www.eere.energy.gov*.
- The **Energy Star®** program is sponsored by the EPA. Energy Star®-qualified new homes are certified to meet strict guidelines for energy efficiency, and certification is verified by an independent third party. All Energy Star®-qualified new homes receive the Energy Star® label, indicating that these energy-efficient homes are more comfortable, more durable, less expensive to own, and good for the environment. More information can be found on their website at *www.energystar.gov*.
- **National Fenestration Rating Council (NFRC)** is a non-profit, public/private organization created by the window, door, and skylight industry. It is comprised of manufacturers, suppliers, builders, architects and designers, specifiers, code officials, utilities, and government agencies. NFRC provides consistent ratings on window, door, and skylight products. More information can be found on their website at *www.nfrc.org*.

Review Questions

1. A goal of green building is:
   a. Produce less pollution
   b. Use different types of energy
   c. Consume more raw materials
   d. Use all natural materials

2. Green buildings help to reduce:
   a. Electrical energy demands
   b. Water usage
   c. Atmospheric pollution
   d. All of the above

3. A life cycle analysis includes all of the following, *except*:
   a. Assessment of raw materials
   b. The type of building
   c. Use of materials
   d. Disposal of materials

4. When peak oil is reached, the following happens:
   a. The rate of oil production enters terminal decline
   b. Oil supplies run out
   c. The maximum rate of oil extraction is reached
   d. a and c

5. Increased CO_2 concentrations in the atmosphere may result in which of the following:
   a. Decrease in global surface temperatures
   b. Increased global surface temperatures
   c. No change in the temperature
   d. Improved air quality
   e. Transit-oriented development

6. Transit-oriented development is centered on the following:
   a. The automobile
   b. Trees
   c. Transit systems
   d. Bicycles

7. Transit-oriented development is an attempt to:
   a. Encourage urban sprawl
   b. Reduce dependence on personal vehicles
   c. Reduce pollution
   d. All of the above
   e. b and c only

8. The Health House® building program is currently the only residential building program that combines which of the following characteristics:
   a. Health
   b. Energy
   c. Resource efficiency
   d. All of the above

9. All of the following are national certification programs for green construction with the exception of:
   a. Building Green
   b. Energy Star®
   c. National Fenestration Rating Council
   d. Building America

10. Which of the following programs is meant to help individuals make environmentally friendly decisions when it comes to buildings?
    a. Building Green
    b. Whole Building Design
    c. Greenerbuilding
    d. All of the above

INDOOR AIR QUALITY

Indoor air quality is an important component in Leadership in Energy and Environmental Design, or LEED, a USGBC rating system. Several building materials, practices, and systems impact the quality of the indoor air. When preparing for the examination, you should review these materials, practices, and systems carefully, being certain to comprehend the mechanisms associated with their impacts. The following section identifies materials, practices, and systems associated with indoor air quality.

To ensure that the indoor air quality is acceptable for occupants after construction a contractor should conduct a complete building flushout with filtered 100% outside air. Alternative methods such as baking out the VOCs in the building by increasing the temperature inside for several days are not acceptable. Also, a complete air testing of the indoor air can be completed to ensure air quality. This is *not* a LEED requirement, however it is an EPA recommendation and will improve the air quality.

Building Materials that Impact Indoor Air Quality

Particleboard and most other standard laminated products use binding material that contains formaldehyde in different forms. Formaldehyde is a known carcinogen and lung irritant that can impact the quality of indoor air. The most common types of formaldehyde-based binders are urea formaldehyde, urea formaldehyde with additives, and phenol formaldehyde. Wood products that use binders composed of urea formaldehyde are a large contributor to formaldehyde emissions in buildings.

This occurs because the urea formaldehyde continues to cure throughout its life. This long duration of curing causes the continued offgassing of formaldehyde into the indoor air environment after installation. Phenol formaldehyde is also used as a binder material in composite wood products. Phenol formaldehyde products offgas 90% less formaldehyde emissions than urea formaldehyde because they cure more rapidly.

Ways to Improve Air Quality in Buildings and Homes

There are basic strategies to improve indoor air quality in both commercial and residential applications. They include controlling the source of the pollution, improving the ventilation in the building or home, and filtering or cleaning the air that enters the indoor environment.

Some sources of indoor air pollution can be easily controlled. These include construction materials, pesticides, and others that are introduced by the occupants, such as tobacco smoke and cleaning products. By limiting the amount of pollutants in the indoor environment, air quality will improve.

To reduce the concentrations of indoor air pollutants easily, contractors and homeowners must increase the amount of outdoor air entering the indoors. Opening windows, using fans, or using ventilation systems introduces fresh outdoor air that most air conditioning and heating systems will not. Installing exhaust fans that remove pollutants directly from the air instead of circulating them through the air will also reduce indoor air pollutants.

Filtering or cleaning the air will also improve the indoor air quality. This may be accomplished by a small table-top air filter or with a larger whole-house system. Some air cleaners remove particles in the air, but most are not designed to remove gaseous pollutants. It is important to know what type of pollutant must be eliminated when determining how to improve the air quality.

Mold Growth Prevention

Molds are microscopic organisms belonging to the fungi kingdom that are found virtually everywhere both indoors and outdoors. Mold is the source of many human allergies and may be detrimental to the health of building occupants. Mold requires oxygen, adequate temperature, a food source, and water to grow. The temperature must be warm and the food source can be things like paper, wallpaper, wallboard, sugars (such as fruits and vegetables), fabrics, wood, or dust. The presence of liquid water is not required for the growth of mold. Relative humidity levels in the air of 70%–100% are sufficient for growth. To prevent the growth of mold inside a building, maintain the level of air humidity inside to below 50% and construct and maintain the building envelope to prevent moisture infiltration.

Multiple Chemical Sensitivity

Multiple chemical sensitivity is defined as a sensitivity or allergy-like reaction to many different kinds of pollutants and irritants. The cause of the sensitivity is not well understood and the reactions are different from patient to patient. Because there is no known cause of the sensitivity and the mechanisms are not well understood by the medical community, treatment is difficult. Most people with the sensitivity must avoid the pollutant that causes their reaction. Typical causes are smoke, VOCs, chemicals, petrol fumes, dust, mold, and even pet dander.

Comfortable Indoor Air Environment

Evaporative cooling occurs when liquid is evaporated into the surrounding air and cools an object or a liquid in contact with it. Evaporative coolers are an effective way to maintain the indoor comfort and use less energy than traditional air conditioning systems. They remove humidity from the air and create a more comfortable environment. Evaporative cooling works well in climates where the air is hot and the humidity is low.

Stack effect is the movement of air into and out of buildings through chimneys, flue gas stacks, or other similar structures. The air movement is caused by differences in the indoor-to-outdoor air density. The difference in density is because of temperature and moisture differences. The stack effect is also referred to as the chimney effect.

Backdrafting occurs due to differences in air pressures. Most newly constructed homes operate at low air pressures. Sometimes the air pressure inside is lower than the air pressure outside. If the difference is stronger than the chimney draft, the flue will work backwards. The chimney will become an entry point for outside air. This will push gases such as carbon monoxide into the house.

Natural Air Changes per Hour

ASHRAE Standard 90.2 recommends that houses have a minimum of 0.35 air change per hour of outdoor air to provide acceptable indoor air quality. According to Green Advantage, that rate is also 0.35 or greater. To ensure the quality of the indoor air, the fresh, outside, or make-up air should avoid locations such as garages, sanitary vents, or garages where it may pull air that is contaminated with pollutants.

Prevention of Carbon Monoxide Migration

Carbon monoxide is a dangerous indoor air pollutant that, if left undetected, can cause death to building occupants. Potential sources of carbon monoxide in commercial buildings and homes include combustion furnaces, boilers, water heaters, automobile exhaust, duct leakage, and improper ventilation of fireplaces or wood-burning stoves. To prevent carbon monoxide from entering the occupied space, sealed combustion furnaces, boilers, and water heaters are critical to maintaining a healthy indoor air environment. Additionally being certain that fireplaces are vented correctly and that ducts are sealed will help minimize the potential for carbon monoxide migration.

Duct Leakage

Duct leakage wastes energy and money as hot or cool air escapes the ducts and does not heat or cool the indoor environment. This leakage can be caused by cracks or loose duct connections. Leaks may also expose individuals to indoor air quality problems. Ventilation systems continually recirculate air throughout a building or home. Leaks in the duct system can draw dust, fumes, pollen, and other contaminants into the ventilation system, impacting the indoor air quality. Use mastic as opposed to tape for duct seams and use insulated ducts with no added formaldehyde.

Testing for Building Air Tightness

A blower door is a tool designed to measure the air tightness of buildings and homes for compliance with energy efficiency standards. A blower door can also help locate air leakage sites that may be the source of wasted energy.

The typical blower door system includes a calibrated fan, a door-panel system, and a device to measure fan flow and building pressure. The operation of a blower door is simple. The calibrated fan is sealed temporarily into an exterior doorframe and blows air into or out of the building or home. The air generated causes a pressure difference between the outdoor and indoor air. The difference in pressure forces air through all holes and penetrations in the building envelope. The more airtight the building, the less air is needed from the blower-door fan to create a change in building pressure.

Thermal Bridging

A thermal bridge is created when there are temperature gradients through the exterior and interior of a building. This can happen through a window frame or actual wall framing material. It typically only occurs in components of a wall system that are exposed on both sides to indoor and outdoor air, like a window frame. Thermal bridging through the building envelope can increase the load on mechanical systems. Heat is transferred through dense material to the colder side of the wall or system.

Insulation

R-values are a measure of the ability of a material to resist the movement of heat. The higher the R-value, the greater the insulation provided by the material. R-values are given in terms of values per inch of thickness of the material. The R-values per inch of thickness of some common construction insulation materials, as well as some alternative insulating materials are listed below. These R-values were retrieved from various insulation manufacturer websites, textbooks, and alternative building product websites. An excellent source for insulation information is the Department of Energy's Insulation Fact Sheet which can be found at *www.ornl.gov*.

Material	R-value per inch
Fiberglass loose-fill	2.5
Fiberglass rigid panel	2.5
Fiberglass batts	3.1
High-density fiberglass batts	3.6
Wood chips and other loose-fill wood products	1
Wood panels, such as sheathing	2.5
Molded expanded polystyrene (EPS) low-density	3.7
Molded expanded polystyrene (EPS) high-density	4
Extruded expanded polystyrene (XPS) low-density	3.6
Extruded expanded polystyrene (XPS) high-density	5
Cementitious foam	2
Cellulose loose-fill	3
Cellulose wet-spray	3
Open-cell polyurethane spray foam	3.6
Closed-cell polyurethane spray foam	5.5
Straw bale	1.45
Cotton batts (bluejean insulation)	3.7

Review Questions

1. Air testing prior to occupancy is:
   a. a LEED requirement
   b. an EPA recommendation
   c. not required
   d. rarely necessary

2. When flushing a building prior to occupancy, the air used should be comprised of what percentage of outdoor air?
   a. 70%
   b. 80%
   c. 90%
   d. 100%

3. Which of the following is/are true with regard to urea formaldehyde?
   a. It is a binder in particleboard
   b. It is a known carcinogen
   c. It continues to cure throughout its lifetime
   d. All of the above

4. Composite wood products that use phenol-formaldehyde result in _____ less formaldehyde emissions.
   a. 50%
   b. 70%
   c. 90%
   d. 100%

5. All of the following are methods of improving indoor air quality, *except*:
   a. Source control
   b. Improved ventilation
   c. Air cleaners
   d. Tight windows

6. A cost-efficient approach to protecting indoor air quality that does not increase energy costs is:
   a. New ductwork
   b. Source control
   c. Increased ventilation
   d. Air filtration system

7. Mold requires what to grow?
   a. Adequate temperature
   b. Nutrients
   c. Moisture
   d. All of the above

8. Multiple chemical sensitivity is a severe sensitivity to the following pollutants:
   a. Chemicals
   b. Petrol Fumes
   c. Smoke
   d. All of the above

9. Evaporative coolers are an effective way to maintain indoor air comfort. When compared to traditional air conditioning systems they:
   a. Use more energy
   b. Use less energy
   c. Remove humidity
   d. Produce more pollution

10. The stack effect helps to drive natural ventilation and infiltration. It is caused by differences in air:
    a. Flow
    b. Temperature
    c. Moisture
    d. b and c
    e. a and b

11. Backdrafting is due to difference in air pressures between indoor and outdoor. This effect allows for:
    a. Gases to enter into the home
    b. Increasing levels of carbon monoxide
    c. More fresh air
    d. b and c
    e. a and b

12. According to Green Advantage, the recommended natural air change per hour is:
    a. 0.20
    b. 0.25
    c. 0.30
    d. 0.35

13. Duct leakage impacts all of the following, *except*:
    a. Outdoor air quality
    b. Indoor air quality
    c. Cooling efficiency
    d. Heating efficiency

14. Duct leakage can draw the following into a home:
    a. Dust
    b. Pollutants
    c. Fumes
    d. All of the above

15. A blower door is a diagnostic tool designed to measure:
    a. The air tightness of buildings
    b. The air flow through a building
    c. Air leaks
    d. a and c

16. Thermal bridging can increase the load on the mechanical systems of a building and results from:
    a. Components in the exterior wall system
    b. Window frames
    d. Interior walls
    d. a and b

17. The R-value of a material is an indication of:
    a. How well it resists the movement of heat
    b. The radius of the material
    c. The thickness of the material
    d. Indicates the type of construction in which it can be used

18. High-density extruded expanded polystyrene has a greater R-value than:
    a. Fiberglass loose-fill
    b. Straw bale
    c. Cotton batts
    d. All of the above

POLLUTION CONTROL AND MEASURES

Environmental pollution takes many forms and there are many ways to control pollution. This next section will present forms of pollution that are addressed in the Green Advantage material as well as ways to prevent and control these pollution sources or alternate sources that produce less pollution.

Light Pollution

Light pollution is a form of pollution whose source is misdirected or misused light. The primary source of light pollution is the inappropriate application of outdoor lighting. It is one of the easiest pollutions to clean up and prevent. Reducing light pollution can be accomplished in several ways, some of which also save energy. Some steps to reducing light pollution include lighting only what is required to the level that is needed, using full cutoff light fixtures that only shine light on the ground, and shading existing outdoor light fixtures. Other cost-saving measures include installing motion sensors to control outdoor lights or using reflectors instead of light fixtures.

Biofuels

A biofuel is defined as solid, liquid, or gas fuel derived from biological material. Biofuels may be produced from any biological source. Biofuels are able to be produced from crops high in sugar like sugar cane or sugar beet, or from plants that are rich in vegetable oil, such as soybeans. Ethanol is produced from materials high in sugar and is a controversial fuel type since a lot of petroleum is used for its production. Oils whose viscosity can be decreased and burned in diesel engines are produced from the vegetable oil rich plants. A new trend in biodiesel production involves the processing of used vegetable oil, as well as wood. Wood can be converted to either methanol or ethanol fuel.

Living Roofs

Living or green roofs utilize living plants and soil on top of a building in order to absorb, collect, filter, and reuse rainwater while preventing run-off. In addition to preventing run-off, these roofs also serve to reduce the heat island effect, reduce the building heating and cooling costs, and filter carbon dioxide out of the air. Living roofs also last longer than conventional roof systems if they are maintained properly.

Review Questions

1. All of the following are true regarding light pollution, *except*:
   a. It is one of the easiest forms of pollution to clean up
   b. It can be prevented by shading outdoor fixtures
   c. Correcting light pollution can also save energy
   d. It is more likely to be a problem where motion sensors are installed

2. Biofuels can be derived from the following:
   a. Sugar cane
   b. Soybean
   c. Used motor oil
   d. All of the above
   e. a and b only

3. Living roofs benefit the environment in all of the following ways, *except*:
   a. Reducing the heat island effect
   b. Filtering pollutants out of rainwater
   c. Increasing cooling loads
   d. Filtering CO_2 out of the air

HEATING AND COOLING

Heating and cooling of buildings requires large amounts of energy and is a great source of natural resource consumption and pollution. The following section highlights the science behind heat movement, ways to improve the energy efficiency of heating and cooling systems, and alternative methods to heat and cool buildings.

Heat Transfer

The science of heat transfer is critical in the design of effective heating, cooling, and insulation systems that are energy efficient in the home or commercial building. Heat will move from an area of higher temperature to an area of lower pressure. It is the difference in temperature that causes heat to flow. The modes of heat transfer include conduction, convection, and radiation. Conduction occurs when there is a difference in temperature that moves through a material. Construction materials that conduct heat well are referred to as conductors and those that do not conduct heat well are called insulators. Metals are good conductors and wood and fiberglass are good insulators. As discussed earlier, insulation is used to prevent or slow heat from moving from one place to another. Convection occurs when heat moves as a result of temperature increases and causes a mixing of materials. Radiation describes the process of heat moving in waves, specifically electromagnetic waves.

The best and most efficient way to transfer heat is by the process of convection, followed by conduction, then radiation. The less efficient a material is at transferring heat, the better it is at insulating.

Windows Facing South and Longest Building Access on North and South

The location of a building on a site and the orientation of the structure can help to maximize the energy efficiency. The longest wall of the building should face within 15 degrees of true south to receive the most winter solar heat gain and reduce summer cooling costs. Minimizing east and west facing walls from the most intense morning and evening sun and reducing the number of windows reduces excessive summer heat gain.

Minimizing Solar Heat Gain

Reducing the solar heat gain into a building helps to reduce the costs associated with cooling the interior and may reduce the cooling load of the building. There are many ways to reduce the solar heat gain into a building. Some of these are natural methods, while others involve installation of awnings and overhangs (solar shades). Common methods of reducing heat gain include window shading, installing trellises, and utilizing deciduous trees on the south side and coniferous trees on the east and west sides. The largest source of solar heat gain is through windows.

Since the windows are the major source of solar heat gain, shading of the windows is an easy way to reduce the solar heat gain. These shades can be interior to the window, a high-efficiency glazing material, a structural overhang, or eliminating windows all together, being careful not to eliminate all views and natural daylight.

Passive Heating and Cooling

The term *passive* means that energy-consuming mechanical components like pumps and fans are not used in the heating or cooling process of the building.

Passive solar heating uses the natural process of heat transfer to heat a building. The design of buildings that use passive solar heating must carefully consider the site and location of the building, the type of climate, and the materials used for construction. Passive cooling building design uses the principles of physics in the building envelope. These principles are used to slow heat transfer into a building and to remove unwanted heat from a building. This can be accomplished by considering the orientation of the building and including various shading mechanisms to the building exterior such as overhangs, landscaping, cross ventilation, or living roofs.

Hydronic Heating – In-Floor and Baseboard

Hydronic heating is similar to heat from a radiator. A hydronic heating system uses hot water heated by a boiler to heat a space by a combination of radiation and convection. The hot water heated by a boiler is piped to fin-tube baseboard units mounted along walls. The heated air is dispersed by convection as the air rises when it is heated by the baseboard unit. Advantages of this type of heat are that it is energy efficient, quiet, clean, provides even heat, and easily provides for different heating zones. The primary disadvantage of baseboard radiation units is that they provide a slow temperature increase. Another disadvantage associated with these systems is that air conditioning requires a separate ductwork distribution and cooling system. This lack of ductwork does not allow for any air filtration or humidification systems.

With a hydronic floor heating system, hot water circulates through lengths of tubing that is encased in a slab of concrete or joist bays in framed structures and loops back and forth on the subfloor. The tubing that is used for these systems is either polybutylene pipes or synthetic rubber tubing. The major advantages of hydronic systems include the fact that a variety of energy sources can be used to heat the water and that the water retains heat for a longer period of time.

Review Questions

1. The physical mechanisms associated with heat transfer are:
   a. Convection
   b. Conduction
   c. Radiation
   d. All of the above

2. To maximize the energy efficiency of a structure the longest wall of the building should face:
   a. North
   b. South
   c. East
   d. West

3. To reduce solar heat gain, all of the following are effective methods, *except*:
   a. Landscaping
   b. Trellis
   c. Window shading
   d. Facing windows north

4. The majority of solar heat gain comes from:
   a. Landscaping
   b. Doors
   c. Windows
   d. Walls

5. Passive solar heating design considers the following:
   a. Location of the building
   b. The climate
   c. Placement of glazing and shading elements
   d. All of the above

6. Passive cooling of a building can be accomplished by all of the following, *except*:
   a. Air conditioning
   b. Shading
   c. Landscaping
   d. Location of the building

7. Some disadvantages of hydronic heating systems are:
   a. Cooling system and related separate ductwork distribution
   b. No possibility for air filtration or humidification systems
   c. Higher operating costs
   d. a and b

8. In-floor hydronic heating systems circulate water through the floor in:
   a. Copper pipes
   b. Polybutylene pipes
   c. Metal pipes
   d. Concrete pipes

WINDOWS

As mentioned earlier, windows are the primary source of solar heat gain into a building. There are different types of windows and window treatments that can help to reduce these gains. Additionally, there are rating systems that can help contractors determine the performance of windows. These are addressed in the following section. It is recommended that you study an example of the National Fenestration Rating Council label that rates the energy efficiency of a window. This will help to prepare you for the Green Advantage exam.

Low-Emittance Windows

Low-emittance (Low-e) coatings are microscopically thin layers that are placed on window surfaces. They coatings are manufactured from metal or metallic oxide layers. They are transparent to light in the visible spectrum and are virtually invisible. These coatings reduce the U-factor of the window by reducing heat flow that occurs due to radiation. There are different types of Low-e coatings that allow for high solar gain, moderate solar gain, or low solar gain depending on the application.

High-Efficiency Windows

High-efficiency windows provide many benefits to buildings and homes when it comes to energy conservation. In general, windows comprise 10 to 25% of the exterior wall area of homes and buildings. Research studies report that buildings in cold weather climates can attribute up to 25% of their heating load to the windows. In hot weather climates, windows account for up to 50% of the cooling load associated with buildings. High-efficiency windows can reduce these loads and save money through reduced utility bills. In addition to the reduction in energy consumption, these windows provide many other benefits. Some of these include improved occupant comfort, fewer condensation problems, a quieter building, reduced wear on furnishings, and improved indoor air quality.

The National Fenestration Rating Council (NFRC) energy performance label can help to determine how well a window or window system will perform. The functions that are rated on the label indicate how well the window will help to cool a building in the summer, warm a building in the winter, keep out the wind, and resist condensation. The information contained on the label allows builders and consumers to compare one window system with another in a reliable manner. The label is included on windows, doors, and skylights and the ratings indicate how the product performs as a whole system.

The energy performance label lists the manufacturer of the product, describes the product, provides a source for additional information, and includes ratings for one or more energy performance characteristics. The information contained on the label is also available in the NFRC online

Certified Products Directory and can be found on their website at *www.nfrc.org*.

The factors listed on the label are discussed in detail below.

U-Factor

U-factor measures how well a product prevents heat from escaping an assembly, in terms of heat loss. Values range from between 0.20 and 1.20 for most assemblies. Lower U-values indicate a high resistance to the flow of heat and indicate that the assembly has a better insulating value.

Solar Heat Gain Coefficient

The solar heat gain coefficient (SHGC) measures how well an assembly blocks heat caused by sunlight. The amount of solar radiation that is allowed to enter a building through a window is defined by the SHGC. This solar radiation is released as heat inside the building. The values of SHGC range from between 0 and 1. Lower values of SHGC indicate that the window allows less solar heat to enter the building.

Visible Transmittance

Visible transmittance (VT) is a measure of the amount of light that is transmitted through a window assembly. Values of VT range between 0 and 1. Higher values of VT indicate that greater quantities of light are transmitted into the building.

Air Leakage

Air leakage (AL) is determined by the amount of air that leaks through a window area. This value is expressed as the equivalent cubic feet of air leaking through a square foot of window area (cfm/sq ft). Lower values of AL indicate that smaller quantities of air will enter the building through gaps in the window assembly.

Condensation Resistance

Condensation resistance (CR) measures the ability of an assembly to resist the formation of condensation on the interior surface of that product. The higher the CR rating, the better that product is at resisting condensation formation. The CR is expressed as a number between 0 and 100.

Review Questions

1. Low-e coatings are microscopically thin, virtually invisible, metal or metallic oxide layers deposited on a window or skylight glazing surface primarily to:
   a. Increase the U-factor
   b. Reduce the U-factor
   c. Improve visible light transfer
   d. Increase the heat transmitted

2. Benefits associated with high-efficiency windows include all of the following, *except*:
   a. Reduced condensation problems
   b. Increased air penetration
   c. Improved indoor air quality
   d. Reduced wear on furnishings

3. The National Fenestration Rating Council's energy performance rating label can help determine the following about a window:
   a. Heating and cooling rating
   b. Condensation rating
   c. Air leakage
   d. All of the above

ALTERNATIVE CONCRETE MATERIALS

There are many waste products and natural organic materials that are being researched for use in the production of concrete. These include municipal solid wastes, various pozzolan materials, and other recycled alternatives. All of these concepts are highlighted in the material below.

Recycled Materials in Concrete

Portland cement concrete (PCC) effectively uses recycled materials in its production. These materials include the byproducts of other industries, and waste concrete itself. The types of waste materials used in concrete include wastes from manufacturing processes and various types of municipal solid waste. Research is being conducted on using these types of waste products in concrete for different purposes. Specific materials that are included in the research include plastic, glass, discarded rubber tires, steel, and coal combustion byproducts.

Pozzolans

Pozzolan materials can be added to Portland cement to change the properties of concrete or as a replacement for a quantity of Portland cement. Some common additives are granulated blast furnace slag, silica fume, and fly ash. An alternative pozzolan that is being investigated is rice hull ash.

Ground granulated blast furnace slag is obtained by quenching molten iron slag from a blast furnace in water or steam as part of the steel production process. This slag can improve the durability of concrete and can help to extend the lifespan of buildings.

Silica fume is a by-product of the reduction of quartz with coke in electric arc furnaces. Silica fume is added to concrete to improve the properties of compressive strength, bond strength, and abrasion resistance.

Fly ash is a fine powder created by coal-fired electric power generation. Each year power plants in the United States produce quantities of fly ash on the order of magnitude of millions of tons. Fly ash is used as a replacement for different quantities of Portland cement in concrete. When used in concrete, fly ash improves many properties associated with strength and durability. It also helps to minimize aggregate segregation and improve the ease of pumping of the concrete. Fly ash is also used as an ingredient in other construction materials such as brick or masonry block. Concrete produced using fly ash is denser and results in a smoother surface.

Cement Production

Concrete is typically composed of gravel, sand, water, and Portland cement. It is an extremely versatile building material that is used extensively worldwide. Unfortunately, significant environmental problems result from the manufacture of Portland cement. Worldwide, the manufacture of Portland cement accounts for 6–7% of the total carbon dioxide (CO_2) produced by humans, adding the greenhouse gas equivalent of 330 million cars driving 12,500 miles per year.

High-Volume Fly Ash Concrete

High-volume fly ash concrete uses fly ash as a substitute for large portions of Portland cement in concrete mixes. The addition of high volumes of fly ash improves the environmental characteristics of the concrete. To be considered high volume fly ash concrete, the fly ash must replace more than 25% of the Portland cement in concrete. The fly ash creates a stronger, more durable concrete product, because it requires a lower water content and therefore cracking is reduced. Overall, high volume fly ash concrete is less energy intensive to manufacture, has a higher ultimate strength, is more durable, requires less water, uses a waste by-product and creates fewer global warming gases when compared to typical concrete mixes.

Review Question

1. Recycled materials that are being considered for use in Portland cement concrete include:
   a. Plastics
   b. Coal combustion products
   c. Wood
   d. a and b

2. All of the following are pozzolans that have been used in Portland cement concrete, *except*:
   a. Plastic
   b. Fly ash
   c. Silica fume
   d. Rice hull ash

3. Fly ash is a replacement for Portland cement in concrete that:
   a. Results in a lower strength product
   b. Reduces the durability of the concrete
   c. Helps to minimize aggregate separation
   d. Is more expensive than Portland cement

4. Cement production accounts for what percent of the total carbon dioxide produced by humans?
   a. 2–3%
   b. 5–6%
   c. 6–7%
   d. 8–9%

5. How does high-volume fly ash concrete compare to regular concrete mixes?
   a. It requires less water
   b. It is more energy intensive to manufacture
   c. It is less environmentally friendly
   d. It is more likely to crack

RECYCLED STEEL

Steel is the world's most recycled material. In the United States, almost 73 million tons of steel were recycled in 2006. Recycling is completed for economic as well as environmental reasons. The recycling of steel is always cheaper than mining virgin ore and moving it through the process of making new steel. Many construction products contain various quantities of recycled steel. These materials and their content of recycled steel are discussed below.

Recycled Content in Steel

Steel is produced through heating raw ore or old steel in furnaces. The basic oxygen furnace (BOF) method of steel production incorporates 25–35% recycled steel in the process. The products produced with this steel include storage containers and automobile fenders. The electric arc furnace (EAF) method of steel production typically incorporates over 80% recycled steel in the process. The EAF process produces products such as structural beams, steel plates, and reinforcement bars. The recycled content found in light gauge steel is less than that in structural steel. Light gauge steel contains about 35% recycled content.

Review Questions

1. All of the following are true with regard to recycled steel, *except*:
   a. It is the most recycled material in the world
   b. Recycling is always cheaper than mining virgin ore
   c. Few construction materials contain recycled steel
   d. The BOF method incorporates up to 35% of recycled steel

2. The recycled content of structural steel produced using the EAF method is approximately:
   a. 60%
   b. 70%
   c. 80%
   d. 90%

FLOORING

There are several affordable and durable options made from grasses and trees that mature in half of the time that it takes hardwoods to reach market size. Sustainable alternatives to traditional hardwood floors include materials such as cork or bamboo flooring products. Sisal flooring is carpeting made from sisal, a succulent plant, and is a popular alternative to traditional carpeting. The various types of flooring materials that are becoming popular alternatives to hardwood are discussed below.

Cork Flooring

Cork flooring is produced from the bark of the tree only. Removing of the bark does not harm the cork tree. The bark grows back without any degradation to the integrity of the tree. The cork bark harvesting occurs every nine years. Cork flooring is extremely resilient and quiet, because 50% of the volume of material is air. This air acts as a natural shock and sound absorber and contributes to the excellent insulating capacity of the material. Floors made of cork stay warm even when installed over cold concrete floors and they are a great covering over radiant floor heating.

Linoleum Flooring

Natural linoleum is a durable resilient flooring product made from linseed oil, pine resin, wood flour, cork powder, limestone dust, natural pigments, and jute. Natural linoleum can be used anywhere a resilient surface is required to include flooring, countertops, and desktops. Natural linoleum is more expensive than vinyl flooring, but due to its environmental and performance advantages over vinyl flooring, it is becoming more popular with consumers.

Vinyl Flooring

Vinyl is a synthetic material. The main ingredients of vinyl are salt and fossil fuel. The manufacturing of vinyl flooring produces VOCs at different stages in the production. Chemicals such as plasticizers and stabilizers are added to tiles and sheet goods to give them flexibility, improve strength, and prevent vinyl from degrading in UV light or heat. Vinyl is one of the most efficient plastics produced in terms of the energy use because 99% of the vinyl becomes part of the finished product.

Review Questions

1. The following are all considered environmentally friendly flooring materials, *except*:
   a. Cork
   b. Linoleum
   c. Hardwood
   d. Sisal

2. The process for creating vinyl flooring:
   a. Is highly energy efficient
   b. Uses salt and fossil fuels
   c. Produces VOCs
   d. All of the above
   e. a and c only

WATER-SAVING MEASURES

There are many ways to save water in both residential and commercial applications. Rainwater harvesting and making more water efficient choices in bathroom fixtures and other landscaping systems will help to save this valuable resource.

Rainwater Harvesting

Rainwater harvesting is the gathering, or accumulating and storing, of rainwater. With added filters, the collected water can be used for drinking water, domestic water, water for livestock, and as water for small irrigation purposes.

To collect the rainwater, a catchment system must be designed. The system is founded upon the principles of gravity. The process of collection occurs in several steps that each have their own components. First, rainwater runs off the roof of the house into rain gutters. The water collected is moved along the gutters and then emptied into a standpipe. When the standpipe has reached its capacity, the overflow water runs through a pipe that empties into a storage drum. Typical systems have two barrels that collect the overflow from the storage drum. Spigots are installed in the bottom of the standpipe and the storage drum. The pressure of the stored water in the standpipe and the drum pushes the water out when the spigots are opened. Commercial and agricultural applications often use larger in-ground systems. These may be transferred to usage areas with solar-powered pumps.

The condition of the rainwater catchment roof, gutters, pipes, and tank are very important in the entire process. These components directly affect the potential for contamination of the water supply. This is critical if the water is to be used for drinking. Contaminants in the water such as bacteria, nitrate, and toxins can be harmful and even fatal. It is critical that all components of the system are maintained to eliminate the potential for contamination.

Non-Potable Water

Non-potable water sources include rainwater, graywater, and municipal recycled water. Water from clothes washers, bathing facilities, and bathroom faucets is referred to as graywater. Water from toilets, kitchen sinks, and dishwashers is not considered graywater due to the various contaminants it contains. It is called blackwater. Municipal recycled water (not potable) includes water from fire hydrants and residential water services.

Xeriscaping

Xeriscaping is the selection of native plants that require little to no irrigation after they are established in the soil. Xeriscaping can be implemented in any location. Effective xeriscaping involves the careful preparation of the soil in which the native plants will be placed. Well-prepared soil absorbs and holds moisture and encourages plants to grow deep roots. With deep roots, plants can access moisture when topsoil is dry and are able to handle low-water conditions. Other key principles associated with the concept of xeriscaping include capturing rainwater for use in watering plants, using mulches, and maintaining the landscaping.

Dual-Flush Toilets

A dual-flush toilet is a high-efficiency, water-saving toilet. A dual-flush toilet has two buttons that can be used for flushing. The first button is for liquid waste. This button uses half of the water that a normal flush would simply because urine is partly water and does not take much work to flush. The typical dual-flush toilet uses 0.8 gallons per flush for liquid waste.

The second button is for solid waste. Solid waste requires more water to flush. The flush operated by the second button is still using less water than in an average single flush. Each flush of the traditional household toilet (prior to new laws) uses on average 2.9 gallons of water. The second button's flush for a dual-flush toilet uses only 1.6 gallons per flush.

Low-Flow Shower Heads

There are opportunities to reduce water consumption and save money by installing low-flow shower heads. According to the EPA, about 33% of indoor water use in the home is associated with bathing. Conventional shower heads have a flow rate between 3–5 gallons per minute, while low flow shower heads have a flow rate of 1.5–2.5 gallons per minute. These save water, energy, and money.

Energy Saving Hot-Water Systems

Saving energy used for hot-water heating can be accomplished by using tankless or on-demand water heaters or an on-demand hot-water recirculation system. These systems are more energy efficient than conventional continuous hot-water recirculation systems. Tankless water heaters heat the water as it flows through the heater, and do not store any water internally except for what is in the heat exchanger coil. In addition to saving energy, these water heaters provide a continuous flow of hot water, compared to a limited flow of continuously heating hot water from conventional tank water heaters. Solar hot water systems and super insulated hot water tanks can also be used.

Review Questions

1. A rainwater catchment system is based upon the collection of rainwater and requires:
   a. A clean roof
   b. Gravity flow
   c. an electric pumping system
   d. b and c
   e. a and b

2. Rainwater harvesting consists of all of the following, *except*:
   a. Transporting
   b. Treatment
   c. Gathering
   d. Storing

3. Municipal recycled water includes water from:
   a. Lakes
   b. Residential water services
   c. Public toilets
   d. Rivers

4. Well-prepared soil can help to conserve water by:
   a. Encouraging deep root growth
   b. Encouraging shallow root growth
   c. Absorbing water
   d. a and c

5. Xeriscaping embodies the following principles:
   a. Using appropriate plants
   b. Capturing rain run-off
   c. Using mulches
   d. All of the above

6. Dual-flush toilets:
   a. Require two flushes
   b. Have two flushing options
   c. Use more water
   d. None of the above

7. Low-flow shower head have flow rates of:
   a. 1.5–2.5 gpm
   b. 2.5–3.5 gpm
   c. 3.5–4.5 gpm
   d. 4.5–5.5 gpm

8. Tankless water heaters:
   a. Are less efficient than continuous hot-water circulation heaters
   b. Are less expensive than continuous hot-water circulation heaters
   c. Operate on demand, as opposed to continuously
   d. Are more common than continuous hot-water circulation heaters

ALTERNATIVE ENERGY SOURCES

In an effort to reduce greenhouse gases and reduce the carbon footprint associated with the built environment, there are many alternative energy sources to fossil fuels being developed. Each of these alternatives has advantages and disadvantages. When preparing for the Green Advantage test, it is highly recommended that you comprehend these advantages and disadvantages and their associated energy source.

Units of Thermal Heat Energy

A unit of thermal heat energy is a BTU, short for British Thermal Unit. The value of a single BTU is defined in terms of the amount of energy needed to heat one pound of water one degree Fahrenheit. To better understand this concept, think of a BTU in light of the following example. If 16 ounces of water (the equivalent of a pound) at 65°F is placed into a pot on a stove and heated, it would take one BTU to raise the temperature of the water to 66°F.

A BTU is approximately a third of a watt of electrical power. The reference above highlighted the energy used to heat a pound of water, but BTUs are also used in reference to cooling. The BTU rating associated with an air conditioning system refers to the amount of thermal energy that is removed from an area. This means that a 55,000 BTU heater and a 55,000 BTU air conditioner are of roughly the same capacity and size. As the value of the BTU output increases, so does the heating or cooling power of the system.

Energy Demand Reduction Steps

A simple step that can be taken to reduce the energy demand within a building is to examine the plug-load equipment used in the building. When determining the types of equipment and other appliances for a building, look for energy efficient models. Plug-load equipment usually accounts for more than 20% of the electric use in offices and the home. Common home and office plug-load equipment include computers and monitors, televisions, battery chargers, speakers, printers, copiers, fax machines, scanners, vending machines, light fixtures, large coffee machines, space heaters, water coolers, refrigerators, and other appliances.

Other energy demand reduction steps for homes and buildings include designing the structure for the occupants needs, improving the building envelope design, and installing energy efficient heating and cooling systems.

Wind Power

Wind power is the conversion of wind energy into electricity. This conversion is completed by using wind turbines. Wind produces only about 1% of world-wide electricity use, but the use of wind power is growing rapidly. Wind energy is favored by many environmentalists as an alternative to fossil fuels because it is plentiful, renewable, widely distributed, clean, and produces lower greenhouse gas emissions.

Photovoltaic Systems

Photovoltaic systems generate electricity by capturing the energy from the sun. Advantages of photovoltaic systems include the fact that solar power is pollution free and that facilities can operate with little maintenance or intervention after initial setup, lasting as long as 25 years. When connected to the energy grid, solar electric generation can displace the highest cost electricity during times of peak demand, reduce grid loading, and eliminate the need for local battery power for use in times of high local demand.

Some financial disadvantages to these systems include the fact that the initial cost may not cover lifespan savings and solar electricity is often initially more expensive than electricity generated by other sources. Solar electricity is not available at night, which requires a storage or complementary power system. Also solar cells produce direct current (DC), which must be converted to alternating current (AC) when used in distribution grids. The conversion process incurs an energy loss of 4–12%.

Net Metering

Net metering is an electricity policy for consumers who own small, renewable energy facilities. These may include such facilities as wind or solar power. The term *net* is defined as the energy left after metered energy inflows. Under net metering, a system owner receives retail credit for a portion of the electricity generated by their system. As part of the Energy Policy Act of 2005, all public electric utilities are now required to make net metering available upon request to their customers.

Review Questions

1. A BTU is a unit of measure of:
   a. Thermal heat energy
   b. Energy efficiency
   c. Energy loss
   d. Energy cost

2. Plug-load equipment usually accounts for which percentage of the building load?
   a. 20%
   b. 40%
   c. 60%
   d. 80%

3. To reduce energy demand in a building the following steps can be taken:
   a. Reduce the plug load
   b. Improve the building envelope
   c. Design for occupants
   d. All of the above

4. Wind energy produces how much of the world's electricity?
   a. 1%
   b. 2%
   c. 3%
   d. 5%

5. Photovoltaic systems require:
   a. A battery backup
   b. A large amount of maintenance
   c. A convertor from DC to AC
   d. None of the above

6. All of the following are true with regard to net metering, *except*:
   a. It is not commonly available
   b. It is available for small solar polar facilities
   c. It is available for small wind power generators
   d. It provides a retail credit for net energy flow

ALTERNATIVE CONSTRUCTION METHODS AND MATERIALS

Construction Impacts

In addition to the impacts that buildings have on the environment, there are other construction activities that impact sites. These include dumping, spilling of hazardous materials, storage of materials, and water diversion. There are simple ways to mitigate these impacts that can be implemented on any jobsite. They include:

- Designate specific areas for storage and disposal
- Clean up spills immediately
- Use pesticides according to the label
- Channel water away from trees
- Use grass swales

Structural Insulated Panels

Structural insulated panels (SIPs) are high-performance building panels used in floors, walls, and roofs. They are most commonly used for residential and light commercial buildings. The panels are typically made by placing a core of rigid foam plastic insulation between two structural layers of oriented strand board (OSB). SIPs are cost effective and energy efficient, while also exhibiting great structural strength. Building with SIPs saves time, money, and labor. While their cost may be initially more expensive, building with SIPs generally costs about the same as building with wood frame construction when you factor in the labor savings resulting from shorter construction time and less job-site waste. Additional savings associated with the use of SIPs include reduced utility bills because less expensive heating and cooling systems are required.

Insulated Concrete Forms

Insulating concrete forms (ICFs) are rigid plastic foam forms that hold concrete in place during curing and remain in place. After the concrete cures, the forms serve as thermal insulation for the walls.

Alternative Methods for Framing – Optimum Value Engineering (OVE)

Alternative framing techniques are at the center of the OVE system. Implementation of the OVE principles results in lower material and labor costs and improved energy performance within a building. The focus of OVE techniques is basic engineering principles. The principles are used to identify ways to minimize material usage while meeting all structural code requirements and limiting thermal bridging. This reduces material costs and waste in a building. The techniques used include changes to construction methods such as two-stud exterior corners, headers only over loadbearing windows, and the use of engineered wood products instead of virgin lumber.

Alternative Lumber

Medium-density fiberboard (MDF) reduces wood waste. MDF can be produced from 100% recovered wood fiber. This reduces landfill waste, slows deforestation, and preserves animal habitats. Compared to particleboard, MDF is stronger, denser, and has superior resistance to warping that may occur over time. The superior performance of MDF has lead to it being the preferred material for furniture production.

Salvaged Lumber

When utilizing salvaged lumber on a project, it may be expensive because there is a need for an independent lumber grader to certify that the lumber is suitable for the application.

Vapor Barriers

The term *vapor barrier* is often used to refer to any material that resists diffusion of moisture through wall, ceiling, and floor assemblies of buildings. The use of vapor barriers is somewhat outdated in modern construction, but some older building codes still require their use. The use of vapor barriers is a function of climate zone. In climates that require a large amount of heating, the vapor barrier is required on the interior of the building. The opposite is true in climates where cooling dominates. They require an exterior vapor barrier. In mixed climates, it not recommended to have a vapor barrier, as moisture will need to move in and out of the building.

Review Questions:

1. SIPs are energy efficient and cost effective because:
   a. They use less material
   b. They use cheaper material
   c. They use plastic
   d. All of the above

2. OVE techniques result in:
   a. Higher labor costs
   b. Lower labor costs
   c. Structural defects
   d. Visual defects

3. MDF is popular in furniture construction because it:
   a. Uses premium lumber
   b. Uses recycled plastics
   c. Is strong and durable
   d. Uses only hardwoods

4. In mixed climates a vapor barrier is:
   a. Required in exterior walls
   b. Below floors
   c. Not recommended
   d. a and b

"green jobs": a workforce system framework for action

FOUNDATIONAL

Policy Drivers/ Interests	Economic Recovery & Job Creation	Legislation/ Funding	Economic Competitiveness	Energy Independence, Efficiency & Security	Save the Planet	Eco-Equity

Transforming Industrial Sectors & Occupations	Energy Generation/Efficiency/Security				Environmental Protection		
	Renewable/ Green Energy (wind, solar, geo, biomass)	Sustainable Manufacturing	Construction/ Skilled Trades	Transportation	Government Oversight	Water Management	Materials & Waste Management

Skill Changers	New Technologies	New Processes	New Materials

OPERATIONAL

	Align Policies & Programs/ Inclusive Investments/ Service Delivery Strategies	Education & Training Models	Strategic Partnerships for Workforce Solutions	Workforce Information	Knowledge Sharing/ Networking/ Communications	Research/ Accountability
ROLES		Apprenticeship Youth Pipeline Community Colleges	Industry Labor Education at all Levels Federal Partners CBO/FBO	Labor Market Intelligence Skills & Competencies Career Information Adult Education		
ACTIONS	1. Program resources focused on re-skilling/ re-employing in green jobs 2. Program participants have access to "green skills" & "green jobs" 3. Build on & leverage existing sector initiatives & green workforce solutions 4. Program policy & direction & strategic investments reflect focus on "green" 5. One-Stop staff have knowledge about "green jobs" & service delivery strategies have "green" focus	1. Map "green" education pathways leading to portable and industry recognized credentials – short term/ long term 2. Identify and leverage existing education & training resources/ investments 3. Educate & train to industry standards/ credentials 4. Build capacity of community colleges 5. Refine/develop "green" apprenticeship models 6. Prepare workers for emerging industry occupations	1. Engage/convene strategic partners 2. Align "green" strategies with regional economic recovery and growth strategies 3. Define workforce challenges 4. Develop innovative workforce solutions 5. Asset map to share/ leverage knowledge, products, and resources 6. Collaborate! Collaborate! Collaborate!	1. Collaborate to develop & leverage new and existing data/information sources 2. Project new jobs, changing jobs, occupational definitions & new skill needs 3. Refine/develop competency models, assessment tools, & career ladders/ lattices 4. Adapt and refine career guidance tools 5. Collaborate to disseminate	1. Develop new "green" learning opportunities for workforce system staff at all levels 2. Fully leverage the communication, knowledge sharing, & e-learning capacity of Workforce3 One 3. Build & nurture a "green jobs" community of practice 4. Leverage conferences & other communication forums to network & share knowledge 5. Promote peer to peer & mentorship learning opportunities and pre fessional development	1. Research & evaluation 2. Performance reporting/ benchmarking 3. Comprehensive system for training grantees/sub-grantees and monitoring compliance at all levels 4. Competitive procurement for contracts & grants 5. Transparency of grant and contract making to public.

Employment & Training Administration (ETA) - U.S. Department of Labor (www. doleta.gov)

CONTREN® LEARNING SERIES — USER UPDATE

NCCER makes every effort to keep these textbooks up-to-date and free of technical errors. We appreciate your help in this process. If you have an idea for improving this textbook, or if you find an error, a typographical mistake, or an inaccuracy in NCCER's Contren® textbooks, please write us, using this form or a photocopy. Be sure to include the exact module number, page number, a detailed description, and the correction, if applicable. Your input will be brought to the attention of the Technical Review Committee. Thank you for your assistance.

Instructors – If you found that additional materials were necessary in order to teach this module effectively, please let us know so that we may include them in the Equipment/Materials list in the Annotated Instructor's Guide.

Write: Product Development and Revision
National Center for Construction Education and Research
3600 NW 43rd St, Bldg G, Gainesville, FL 32606

Fax: 352-334-0932

E-mail: curriculum@nccer.org

Craft ______________________ Module Name ______________________

Copyright Date ______________ Module Number ______________ Page Number(s) ______________

Description

____

____

____

____

(Optional) Correction

____

____

____

(Optional) Your Name and Address

____

____

____

Your Role in the Green Environment

Index

D

E

F

X